Nikolay Sokolov

Optimization methodology of hierarchical control of orbit groups

Nikolay Sokolov

Optimization methodology of hierarchical control of orbit groups

Basics, Concepts, Methods

LAP LAMBERT Academic Publishing

Impressum / Imprint

Bibliografische Information der Deutschen Nationalbibliothek: Die Deutsche Nationalbibliothek verzeichnet diese Publikation in der Deutschen Nationalbibliografie; detaillierte bibliografische Daten sind im Internet über http://dnb.d-nb.de abrufbar.
Alle in diesem Buch genannten Marken und Produktnamen unterliegen warenzeichen-, marken- oder patentrechtlichem Schutz bzw. sind Warenzeichen oder eingetragene Warenzeichen der jeweiligen Inhaber. Die Wiedergabe von Marken, Produktnamen, Gebrauchsnamen, Handelsnamen, Warenbezeichnungen u.s.w. in diesem Werk berechtigt auch ohne besondere Kennzeichnung nicht zu der Annahme, dass solche Namen im Sinne der Warenzeichen- und Markenschutzgesetzgebung als frei zu betrachten wären und daher von jedermann benutzt werden dürften.

Bibliographic information published by the Deutsche Nationalbibliothek: The Deutsche Nationalbibliothek lists this publication in the Deutsche Nationalbibliografie; detailed bibliographic data are available in the Internet at http://dnb.d-nb.de.
Any brand names and product names mentioned in this book are subject to trademark, brand or patent protection and are trademarks or registered trademarks of their respective holders. The use of brand names, product names, common names, trade names, product descriptions etc. even without a particular marking in this works is in no way to be construed to mean that such names may be regarded as unrestricted in respect of trademark and brand protection legislation and could thus be used by anyone.

Coverbild / Cover image: www.ingimage.com

Verlag / Publisher:
LAP LAMBERT Academic Publishing
ist ein Imprint der / is a trademark of
OmniScriptum GmbH & Co. KG
Heinrich-Böcking-Str. 6-8, 66121 Saarbrücken, Deutschland / Germany
Email: info@lap-publishing.com

Herstellung: siehe letzte Seite /
Printed at: see last page
ISBN: 978-3-659-62083-6

Table of contents

Annotation

The methodological approach is proposed for optimization of hierarchical control of spacecraft orbit constellations. It is based on the methods of optimal control theory and waiting theory. The main hierarchical levels of the control system are determined. The results of optimal control task decisions are presented in the powered phase of orbital injection, during the deorbiting maneuvers, analytical algorithms are described for accelerated determination of orbit parameters of space objects by the measurements of onboard optical sensors. The interaction pattern is developed for separate system components, Kolmogorov differential equations are worked out for definition of probability of finding the system in each of single-out states. Using this methodology it is possible to conduct comparative efficiency evaluation for alternative variants of system structure and determine critical units of the system.

Introduction

In the context of predicted large-scale development of spacecraft (SC) orbit groups, the requirements are toughened for the complex decision of the whole series of interconnected tasks influencing the control process and optimal fulfillment of payload functions. In particular, they include efficient ballistic construction of orbit constellations, safety of SC flights in the near-earth space debris environment, allocation of ground control means and relaying and communication facilities, increasing in operational efficiency update of SC onboard equipment diagnostics, decision-making and implementation, development of efficient algorithms for elimination of contingencies, etc.

At that, providing the high level efficiency of target program fulfillment must be pointed out as the main control task, along with the realization of energy-optimal motion patterns of SC in all flight stages: shaping of satellite orbits; correction actions of orbit parameters constraint, SC avoidance maneuvers from space debris, interorbital transfers, and satellite deorbiting.

One of the basic differences between the control of orbit constellations and control of solitary spacecraft is a considerable dependence of object tasks decision not only on immediate control decision-making but also on the structure and composition selection of SC control system (ground and space control means, hardware and software and communication interfaces, in-flight equipment and algorithms of independent control, etc.). At the same time the functioning of each component is described by means of different-type specific models and this complicates the general mathematical formalization of control by traditional methods.

According to the above-stated there is a need for elaboration of new methodological approaches for optimization of orbit constellation control including the determination of optimal structure and functioning processes of control system. As there are a lot of elements which influence the control process and the variety of multitype control and information ties between them, the control system of the orbit group as an object of high complexity level may be classified as the large hierarchical control system (HCS) [1,2].

A whole number of works is devoted to the problem of hierarchical control systems optimization [1-16]. It should be noted that specific tasks are mainly considered in these works. Thus, the issues of structural optimization of hierarchical systems are studied in the works [1-5]. The modern methods for optimization of system

3

characteristics and their control are described [6]. The conducted research is directed to the structures of high dimension with the formation of hierarchical levels which enable to find an optimal solution with a considerable decrease in computational costs. The structure optimization tasks are considered in the number of works for specific variants of the systems and different solution methods are proposed. For example, the solution is being searched for hierarchical systems of resources allocation to optimize the structure for a certain class of objective functions [7-8]. The branch and bound method is proposed for the task of hierarchical structure optimization represented as a model of discrete programming with a special aim function [9]. The description is given for mathematical model of formation of multilevel waiting system and the results of system optimization under certain conditions are presented [10]. There are the simple but informative examples of hierarchical structures with non-trivial optimal structure [11, 12]. The problem of synthesis of optimal hierarchical structure is considered as a task of minimizing the functional at a set of oriented acyclic graphs. The algorithmic apparatus is developed which describes the large class of tasks, admitting different conceptual interpretation [13]. The structure of information expert system is suggested supporting the modeling tasks of development dynamics and performance analysis of control processes of complex hierarchical systems [14]. In the article [15] the research algorithm of disturbed motion paths of SC in the atmosphere is proposed on the basis of Markov process theory. The SC motion is interpreted as a probability change of its stay in some fixed positions of the phase space. In the work [16] there are the results of comparative analysis of system-building methods of automated SC control systems.

At the same time, the analysis of existing materials showed that there's no general methodology of a complex optimization of hierarchical control systems for satellite constellations control with regard to meeting the generalized requirements for performance of specific tasks decision, minimization of necessary power inputs for the functioning of SC, flight safety of SC in the space debris environment.

The methodological approach to the problem decision of complex optimization of hierarchical spacecraft control is presented in this work on the basis of waiting theory and optimal control theory.

1. The basic principles of methodological approach to the optimization of hierarchical control systems

This work presents the mathematical formalization of processes which take place during the control of spacecraft constellation. The general principles of methodological approach to the search of optimal control decisions are substantiated taking into account the specifics of SC flight dynamics in different flight phases.

The basis of the methodological approach is iterative determination of the rational structural composition of hierarchical control systems and step-by-step decision of optimization tasks with the following conjugation of the obtained results. At the first stage the alternative variants of system structure are considered on the basis of private task decision. On their basis the optimization of function allocation performed by separate elements is realized as well as coordination of aims and optimization interaction of elements of different hierarchical levels. The alternative variants of system structural construction are analyzed at the second stage. The optimal system structure is determined in the result of synthesis of main control functions.

The general regularities of structural construction of hierarchical control systems are implemented during the analysis of the alternative variants:

- successive vertical layout of subsystems, forming the appropriate hierarchy;
- action priority or the right of "interference" of the upper level subsystem in the actions of the lower level subsystems;
- dependence of actions of upper level subsystem on actual function performance by the subsystems of the lower level;
- considerably greater degree of uncertainty in the description of control tasks at the upper levels of hierarchy.

Process analysis of system elements functioning during the control of SC groups reveals the main levels of hierarchy:

- the upper level (coordination level) – realizes the coordination of all actions for system control aiming at the most efficient implementation of object functions by SC groups;
- the second level (self-organization level) – implements the choice of criteria and algorithms used at the lower levels of hierarchy for the purpose of the main control task decision;
- the third level (adaptation level) – realizes concretization of uncertainty sets for the subsystems of the upper levels by means of analysis of current situation and

elaboration of requirements for the solution algorithms by the subsystems of the lower level;

- the fourth level (choice level) – realizes the solutions of specific tasks according to the initial data and algorithmic statements from the upper level systems.

Taking the above-mentioned into account, as applied to the study of hierarchical control optimization of spacecraft constellations, the following solution pattern of tasks is considered and the functional criteria at the dedicated hierarchy levels are specified.

At the coordination level the comparative performance analysis of alternatives of control system construction is conducted, the critical units of the system are revealed, the recommendations are given for the choice of optimal variant providing the most efficient coordination of control functions.

At the self-organization level the tasks of optimal control are solved during the functioning of the space system, and the requirements and criterion for its rational structural characteristics are elaborated. The algorithms are developed for the task decision of necessary power inputs minimization and providing operation speed during SC deorbiting and descent into a specified area of earth and also for maintaining of SC orbital parameters in the prescribed limits.

At the adaptation level the space control tasks in satellite orbits are defined concretely, also in order to provide safe flights in the space debris environment. In particular such tasks include as follows: autonomous orbit determination of potentially hazardous man-made objects, detection of dangerous approaches of SC with space debris, estimated probability of collisions, elaboration of recommendations for conducting avoidance maneuvers of SC.

At the choice level specific tasks are solved for satellite orbit shaping, the algorithms are developed for estimation of necessary power consumption during the SC motion in the powered flight phases.

We shall state the general mathematical interpretation for determination of optimal structure and composition of hierarchical systems. We introduce the symbols:

C_1 – subsystem of the upper level;

C_{2i} $(i = 1, 2, ..., n)$, C_{3j} $(j = 1, 2, ..., m)$, C_{4k} $(k = 1, 2, ..., l)$ – complex of subsystems of the second, the third and the fourth hierarchical levels, respectively;

P – control process;

m – set of control actions;

ω – set of external disturbances;

6

r – set of information actions, generated by the subsystems of HCS;

s – set of coordinating actions;

y – set of output actions;

k – set of information actions, generated by the subsystem of the upper level.

We introduce the functioning model of the HCS in terms of set theory.

The controlled process is represented as: P: $M \times W \to Y$.

The coordinator model is introduced as: $C_1: K \to S$.

The functioning model of s-subsystem ($s = 2, 3, 4$) is realized as: $C_s: S \times R_s \to M_s$.

Accordingly information feedbacks of all levels of hierarchy are represented as:

$f_1: S \times R \times M \to K$; $f_s: M_s \times W \times Y \to R_s$.

The global optimization task reflects the main control purpose and it is stated as follows: it's required to define such set of control actions m, with which the global objective function $g(m) = G[m, P(m)]$ reaches its maximum.

Considering the local optimization tasks, it is assumed that the control process is represented as a composition of subprocesses P_s ($s = 1, 2, 3, 4$) interacting with each other. Moreover, the interaction of each subprocess P_s with others is realized through many links u_s. Let D_s – be the local optimization task solved by s-control body of low level, and the local optimization quality function for solution of this task is as follows: $g_s(m_s, u_s) = G[m_s, P_s(u_s, m_s)]$, where $M_s \times U_s$ is a set of object function values.

There are two main ways to influence the local optimization tasks on the part of high level subsystems:

-   through the quality function G_s by correction of control aims;

-   through changing the variety of links U_s ($s = 1, 2, 3, 4$) in the class of subprocesses P_s.

The first method involves presetting the multitude of local quality functions, as a result of which the coordinating signal S_s is aimed at selecting of appropriate quality function from the specified number of s-control system. In case if there is no possibility of coordination by changing aims, the coordination by changing limits either on the basis of principle of decoupling of interactions U_s or prediction of interactions U_s is used.

The principle of decoupling of interactions implies that each lower-level control body gets the right during solving its own control task to consider linking inputs U_s as additional variables which can be picked out from its own local measures. In this case the solution of control task of the lower level is determined as if the lower-level elements and subprocesses were completely autonomous.

The principal of prediction assumes that the coordinating signals S_s contain information about the predicted values of links S_s which will take place during SC control.

Thus the proposed approach allows forming the multitude of control impacts applied to all hierarchical levels of the system and conducting the comparative analysis of alternative structural composition variants of the HCS in order to determine the optimal variant.

2. Optimal control of spacecraft in the process of satellite orbit shaping. Minimization of necessary power consumption

One of the main problems of SC orbit constellation development is minimizing of fuel consumption necessary for SC ascent to the specified orbits. Due to this the payload share in the SC weight balance can be increased. The rational construction of satellite constellations must provide both reduction of necessary power consumption for maintaining of orbital parameters in the prescribed limits and flight safety improving in conditions of possible collision with space debris. As is known, the areas in the range from 700 up to 900 km are mostly congested. That's why the functioning of SC in these altitudes will be connected with the necessity to quite often conduct the avoidance maneuvers from space debris. On the other hand, the altitude reduction of orbits will result in the growing influence of atmospheric drag on the spacecraft and frequency of corrections for maintaining the necessary flight altitude values.

That is why, it is necessary to solve the SC optimal control tasks in the powered flight phase during the orbit shaping providing the minimum of power consumption.

It should be noted that detecting of spacecraft optimal control in the powered phase of ascent to satellite orbits by the classical methods [18, 19] as opposed to the solution of variational tasks of SC control in the orbital phases and during reentry, is connected with a number of additional complications in the decision of boundary value problems for specification of edge conditions of the flight in the optimal paths and also during the formation of thrust vector control law of the power plant in the analytical form.

Based on the specifics of SC motion with the propulsive, aerodynamic, gravitational, Coriolis and centrifugal forces that affect it, in this work we propose the new methodological approach to the study of SC control problem in the powered flight phase. The application results of the proposed solution are presented in calculation of concrete ascent trajectories.

During the research we paid special attention to the possibility of trajectory calculation in the wide range of orbit altitudes, design and ballistic, weight and power characteristics of SC, boundary conditions, limitations for the maximum allowed values of overloads and air-velocity pressures.

9

The determination of engine thrust vector optimal control in the spacecraft insertion phase is one of the main and complicated tasks solved during the research of SC powered flights. In this phase the SC is flying with the running engines, and acceleration and achievement of specified orbit parameters are provided.

In this work we consider the control problem of one-stage SC during its insertion to the specified orbit with vertical and horizontal takeoffs. As a formal condition of SC vertical takeoff we shall consider the existence of nonzero time span from the flight start Δt, where pitch angle $\vartheta \equiv \pi/2$, for the horizontal takeoff $\vartheta(t\epsilon\Delta t) = 0$.

The motion of SC centre of mass in the velocity coordinate system taking into account Earth's rotation is described with the known system of differential equations, in particular, presented in the work [21],

$$\frac{dV}{dt} = -\frac{\rho V^2}{2P_x} - \frac{\mu}{r^2}\sin\theta + \frac{P}{m}\cos\alpha - \omega^2 r \cos\varphi\,(\sin\varphi\sin\varepsilon\cos\theta - \cos\varphi\sin\theta),$$

$$\frac{d\theta}{dt} = \frac{\rho V K_\sigma}{2P_x}\cos\gamma - \frac{\mu}{r^2 V}\cos\theta + \frac{V}{r}\cos\theta + \frac{P}{mV}\sin\alpha + 2\omega\cos\varphi\cos\varepsilon +$$

$$+ \frac{\omega^2 r}{V}\cos\varphi\,(\sin\varphi\sin\varepsilon\sin\theta + \cos\varphi\cos\theta),$$

$$\frac{d\varepsilon}{dt} = \frac{\rho V K_\sigma}{2P_x\cos\theta}\sin\gamma - \frac{V}{r}\cos\theta\cos\varepsilon\,\mathrm{tg}\,\varphi - \frac{2\omega}{\cos\theta}\cdot$$

$$\cdot\,(\cos\theta\sin\varphi - \sin\theta\sin\varepsilon\cos\varphi) - \frac{\omega^2 r}{V\cos\theta}\sin\varphi\cos\varphi\cos\varepsilon, \qquad (3.1)$$

$$\frac{dh}{dt} = V\sin\theta, \quad \frac{d\lambda}{dt} = \frac{V\cos\theta\cos\varepsilon}{r}\,\frac{1}{\cos\varphi}, \quad \frac{d\varphi}{dt} = \frac{V}{r}\cos\theta\sin\varepsilon, \quad \frac{dm}{dt} = -\frac{P}{P_{spec}g_E},$$

$$P_x = \frac{m}{C_x S}, \quad K_\sigma = \frac{C_y}{C_x}, \quad r = R(1 - \kappa\sin^2\varphi) + h, \quad P = P_V - p_h S_n,$$

where V – spacecraft velocity, θ – burnout angle, ε – angle between projection of velocity onto the local horizon and local parallel, h – spacecraft flight altitude above Earth's surface, λ and φ – geographic longitude and latitude respectively, m – spacecraft mass, R – Earth's equatorial radius, κ – Earth's ellipticity, ρ – atmospheric density, μ – product of gravitational constant by Earth's mass, P_x – front surface reduced load, K_σ – aerodynamic characteristic, C_x and C_y – aerodynamic drag coefficient and lift coefficient respectively, γ – roll angle, ω – earth angular velocity, P – engine thrust, P_V – engine thrust in vacuum, P_{spec} – specific thrust, α – angle between spacecraft velocity vector and thrust vector, g_E – gravitational acceleration on earth's surface, p_h - air pressure at the height h, S_n – nozzle exit shear section.

For determination of velocity V and angles θ and ε in the inertial coordinate system, relations are used [22]:

$$V_i = \sqrt{V^2 + V_{rot}^2 + 2VV_{rot}\cos\theta\cos\varepsilon}, \qquad \theta_i = \arcsin\left(\sin\theta\frac{V}{V_i}\right),$$

$$\varepsilon_i = \arcsin\left(\sin\varepsilon\frac{V\cos\theta}{V_i\cos\theta_i}\right), \qquad V_{rot} = \omega r\cos\varphi.$$

where V_{rot} – Earth's rotation rate.

Dependence of aerodynamic coefficients C_x and C_y on Mach number M and angle α is determined by the equations [23]:

$$C_x = C_{x0} + A(M)C_y^\alpha, \qquad C_y = C_y^\alpha(M)\alpha,$$

where A – dimensionless coefficient, describing SC aerodynamic parameters.

It is supposed that a SC is controlled by engine ignition and cut-off and change of angle α. At that, the values P and α range within:

$$0 \le P \le P_{max}, \qquad -\frac{\pi}{2} \le \alpha \le \frac{\pi}{2}. \qquad (3.2)$$

At launch time $t_0 = 0$ conditions are met:

$$V_0 = V(0), \qquad \theta_0 = \theta(0), \qquad \varepsilon_0 = \varepsilon(0), \qquad h_0 = h(0),$$
$$\lambda_0 = \lambda(0), \qquad \varphi_0 = \varphi(0), \qquad m_0 = m(0). \qquad (3.3)$$

The moment of end of insertion is the terminal time of satellite orbit shaping with the specified apogee altitude h_α and perigee altitude h_π:

$$h_{ak} = h_a(t_k), \qquad h_{\pi k} = h_\pi(t_k), \qquad t_k - not\ specified. \qquad (3.4)$$

At that limitations for the maximum values of velocity pressure are taken into account, as well as longitudinal load factor and the product of velocity pressure values by angle α:

$$q = \frac{\rho V^2}{2} \le q_{max}, \qquad n_x \le n_{xmax}, \qquad q\alpha \le C_{max}. \qquad (3.5)$$

Basically such tasks are solved using the necessary optimality conditions of the Pontryagin maximum principle [18]. At the same time, due to the SC motion dynamics in the powered flight phase and the relevant symbolic model (3.1) on account of formalism of optimality conditions it is impossible to obtain SC control laws in the analytical form.

Write the Hamiltonian:

$$H = -\frac{\rho V^2}{2P_x}\psi_1 + \frac{P}{m}\cos\alpha\,\psi_1 + \frac{\rho V K_\sigma}{2P_x}\cos\gamma\,\psi_2 + \frac{P}{mV}\sin\alpha\,\psi_2 + \frac{\rho V K_\sigma}{2P_x\cos\theta}\sin\gamma + F,$$

where ψ – conjugate variables, F – function, not depending on control parameters.

11

As it follows from analyzing the Hamiltonian, the summands that enter it, characterizing the influence of propulsive and aerodynamic forces, depend on operating angle α. The summands, describing the influence of propulsive forces in an explicit form, include the functions $\cos \alpha$ and $\sin \alpha$. The aerodynamic components indirectly depend on α, because the aerodynamic drag coefficient C_x affects them (3.1). It depends, in its turn, on angle α. In connection with this, angle α optimal control law cannot be obtained in the analytical form. Its determination is connected with the necessity of solution of transcendental equations on each step of integration of differential equations system (3.1). It reduces considerably the processing speed of solution for the variational tasks of such type.

In this work we propose a new method of optimal thrust vector control in the powered flight of spacecraft to satellite orbits. This method consists in the conditional partition of spacecraft path into specific areas, establishing control in each of them, conjunction of obtained solutions and shaping of an interim orbit with apogee altitudes $h'_a \leq h_{ak}$ and $h'_\pi \leq h_{\pi k}$ (or corresponding values V_1, θ_1 and h_1) with the further powered maneuver of spacecraft. Besides, the new injection scenario to satellite orbit is proposed, providing two-burn orbit adjustment upon completion of the powered flight. This injection scenario is based on the shaping of an interim orbit during the spacecraft motion in the powered leg with the parameters $h'_a \leq h_{ak}$ and $h'_\pi \leq h_{\pi k}$ (or corresponding spacecraft speed values V_1, angle θ_1 and height h_1 in the inertial coordinate system at engine cut-off). The use of such injection scenario enables, as a result of spacecraft path calculations with different values of V_1, θ_1 and h_1, to choose the paths with comparatively low mass consumption required for shaping the final orbit.

For both takeoff types the parameters V_1, θ_1 and h_1 ranged within the limits as follows:

$$6.5 \leq V_1 \leq 7.78 \text{ km/s}, \qquad 0 \leq \theta_1 \leq 3°, \qquad 70 \leq h_1 \leq 150 \text{ km}.$$

In case of the vertical takeoff all the powered flight of spacecraft interim orbit injection is divided into three conditional stages: vertical flight, gravity turn and stage of formation of final conditions.

At the first stage the control angle α is determined from the conditions $\dot{\theta} = 0$ and $\theta = \pi/2$. Setting the second differential equation to zero (3.1) we get the dependence for calculation of angle α:

$$\alpha = -\arcsin\left[\frac{mV}{P}\left(\frac{\rho V C_y S}{2m}\cos\gamma + 2\omega\cos\varphi\cos\varepsilon + \frac{\omega^2 r}{V}\cos\varphi\sin\varphi\sin\varepsilon\right)\right].$$

Then the control program for angle α is chosen from the condition $\alpha = -C/q$, as a result of this spacecraft trajectory deviation from the vertical is achieved as well as the turn of spacecraft velocity vector due to gravitational forces. At that, the required level of turn is provided by the choice of coefficient.

At the final stage angle α is determined taking into consideration the assumption of monotone decrease of angle θ during all the powered flight, the sufficient condition for which is as follows:

$$\frac{P}{mV}\sin\alpha \le \frac{\mu}{r^2V}\cos\theta - \frac{V}{r}\cos\theta - \frac{\rho V C_y S}{2m}\cos\gamma - 2\omega\cos\varphi\cos\varepsilon -$$
$$-\frac{\omega^2 r}{V}\cos\varphi\,(\sin\varphi\sin\varepsilon\sin\theta + \cos\varphi\cos\theta).$$

Note that in case of choice of angle α less values, the apogee altitude of a transfer orbit can be less than the chosen h'_α. That's why for calculation of values α we propose a formula:

$$\alpha = arcsin\left\{\frac{mV}{P}\left[\frac{\mu}{r^2V}\cos - \frac{V}{r}\cos\theta - \frac{\rho V C_y S}{2m}\cos\gamma - 2\omega\cos\varphi\cos\varepsilon - \right.\right.$$
$$\left.\left. -\frac{\omega^2 r}{V}\cos\varphi(\sin\varphi\sin\varepsilon\sin\theta + \cos\varphi\cos\theta]\right\} - \frac{\Delta\alpha}{\theta^*}\theta,$$

where $\theta = \theta^*$ - at the end of second stage of flight.

In case of such control $\theta \le 0$ with $\Delta\alpha \ge 0$, and by the relevant choice of value $\Delta\alpha$, the required ascent trajectory slope is provided and the interim orbit is shaped with the given values V_1, θ_1 and h_1 (or h'_α and h'_π).

Then the SC is put into the target orbit by two burns $P = P_{max}$ and by angle $\alpha = 0$ – first at the moment of reaching the parameter values V_1, θ_1 and h_1, and then in the apogee of the transfer orbit. In the first case the engine is shut off when the transfer orbit apogee altitude is raised to the required value of h_{ak}, in the second case when the perigee altitude reaches the required value $h_{\pi k}$. The calculation of current values of spacecraft phase coordinates and active mass is conducted with the use of system of equations (3.1). At that the approximate estimators of reference velocity and usable fuel are obtained by the formulae [24]:

$$\Delta V_1 = \sqrt{\frac{2\mu(r_{ak} - r_1)r_{ak}}{r_1(r_{ak}^2 - r_1^2\cos^2\theta_1)}} - V_1, \quad \Delta V_2 = \sqrt{\frac{\mu}{r_{ak}} - \frac{r_1}{r_{ak}}\cos\theta_1\sqrt{\frac{2\mu(r_{ak} - r_1)r_{ak}}{r_1(r_{ak}^2 - r_1^2\cos^2\theta_1)}}},$$

$$r_1 = R + h_1, \quad r_{ak} = R + h_{ak}, \quad \Delta m = m_1\left[1 - \exp\left(-\frac{\Delta V_1 + \Delta V_2}{P_{spec}g_E}\right)\right].$$

13

The following engine thrust control programs are used in order to provide assumptions for the maximum allowable values of overloads and air-velocity pressures.

In the flight phase of the spacecraft with $n_x = n_{xmax}$ the following relation is performed: $n_{xmax} = V / g_E$, affording with regard to (3.1) get the thrust P equation:

$$n_{xmax} \equiv \frac{V}{g_E} = -\frac{\rho V^2 C_x S}{2 m g_E} - \frac{\mu}{r^2 g_E} \sin \theta + \frac{P}{m g_E} \cos \alpha -$$
$$-\frac{\omega^2 r}{g_E} \cos \varphi \, (\sin \varphi \sin \varepsilon \cos \theta - \cos \varphi \sin \theta).$$

From this equation we can determine the dependence for calculation of thrust values P, which provides SC motion in the phase with the constant overload:

$$P = \frac{m}{\cos \alpha} \left[n_{xmax} g_E + \frac{\rho V^2 C_x S}{2m} + \frac{\mu}{r^2} \sin \theta + \right.$$
$$\left. + \omega^2 r \cos \varphi \, (\sin \varphi \sin \varepsilon \cos \theta - \cos \varphi \sin \theta) \right] .$$

In the flight phase with maximum allowable level of air-velocity pressure the thrust P is calculated subject to conditions:

$$q = \frac{\rho V^2}{2} = \text{const}, \qquad \text{or} \qquad V_{i+1} = V \sqrt{\rho / \rho_{i+1}} \, ,$$

where i - the number of current step of system's integration (3.1).

Equating the required velocity increment $V = V(\sqrt{\rho / \rho_{i+1}} - 1)$ to the approximate value ΔV, taken from the first equation of the system (3.1), we get the equation for determination of P:

$$V \left(\sqrt{\rho / \rho_{i+1}} - 1 \right) = \left[-\frac{\rho V^2}{2 P_x(\alpha)} - \frac{\mu}{r^2} \sin \theta + \frac{P}{m} \cos \alpha - \right.$$
$$-\omega^2 r \cos \varphi \, (\sin \varphi \sin \varepsilon \cos \theta - \cos \varphi \sin \theta)] \Delta t,$$

which is followed by:

$$P = \frac{m}{\cos \alpha} \left[\frac{V}{\Delta t} \left(\sqrt{\frac{\rho}{\rho_{i+1}}} - 1 \right) + \frac{\rho V^2}{2 P_x} + \frac{\mu}{r^2} \sin \theta + \right.$$
$$+\omega^2 r \cos \varphi \, (\sin \varphi \sin \varepsilon \cos \theta - \cos \varphi \sin \theta)],$$

where $\rho_{i+1} = \rho_0 e^{-\beta(h + V \sin \theta \Delta t)}$ - the predicted atmospheric density at the next step of integrating $\Delta t = t_{i+1} - t$.

As a result in case of the vertical takeoff the search problem of control parameters reduces to an iterative task of determination of time values of vertical takeoff t_v, constants C, $\Delta\alpha$ and angle θ^*, interim orbit altitudes h'_α and h'_π, with which the relative minimum of fuel mass consumption is provided as well as the fulfillment of boundary conditions and constraints. The calculation process of problem-solving is drawn up in such a manner, that the final conditions of the preceding phase are the initial conditions of the next one. In case of the horizontal takeoff at the first flight phase the SC injection into a maximum trajectory angle is provided with the predetermined Mach number M_1. It is realized using the following angle α control law:

$$\alpha = arcsin\left\{\frac{mV}{P}\left[\frac{\mu}{r^2V}cos\theta - \frac{V}{r}cos\theta - \frac{\rho V C_y S}{2m}cos\gamma - -2\omega cos\varphi cos\varepsilon - \frac{\omega^2 r}{V}\right]\right\}$$
$$+ A(M - M_0)(M_1 - M).$$

The first summand of this relation corresponds to control by angle α during SC flight with constant angle θ. The second summand is the second order polynomial with respect to M and it becomes zero at the initial instant and when θ_{max} is reached. Choosing the parameter A the required spacecraft trajectory slope is provided in the first phase as well as the prescribed value θ_{max} with $M = M_1$.

The control structure on the second and the third stages is the same as in case of the vertical takeoff. The SC ascent trajectories in case of horizontal takeoff are obtained as a result of parameters determination of A (with θ_{max}), C, $\Delta\alpha$, θ^*, h'_α and h'_π, providing the relative minimum of fuel mass consumption with the fulfillment of specified constraints and boundary conditions.

The proposed method has been tested for the calculation of several concrete ascent trajectories.

On figure 3.1- 3.4 there are the dependencies of velocity V, trajectory angle θ, flight altitude h, SC mass m, thrust P, control angle α, overload n and air-velocity pressure q on the time of motion in the powered insertion phase of SC to the interim satellite orbit.

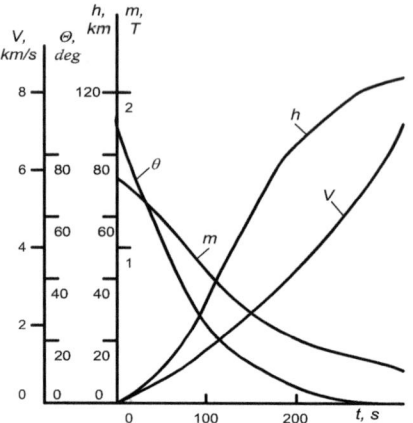

Fig. 3.1. Dependencies of velocity V, angle θ, altitude h, spacecraft mass m on time t in case of spacecraft vertical takeoff. Nominal initial data.

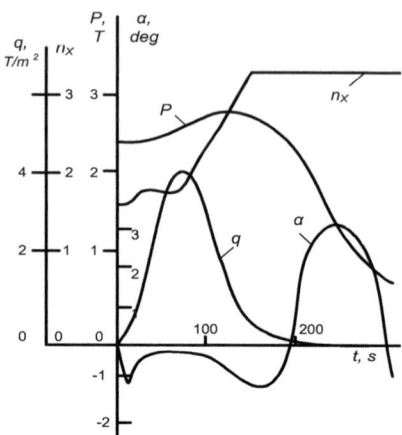

Fig. 3.2. Dependencies of parameters P, α, n_x and q on time t in case of spacecraft vertical takeoff

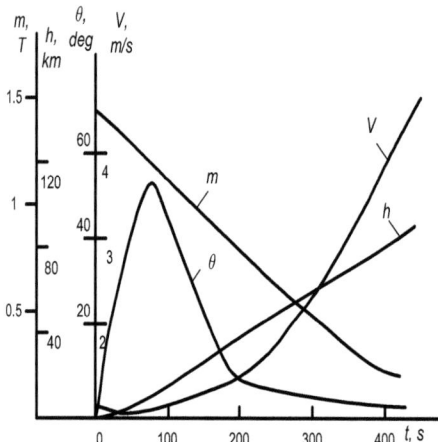

Fig. 3.3. Dependencies of velocity V, angle θ, altitude h, spacecraft mass m on time t in case of spacecraft horizontal takeoff.

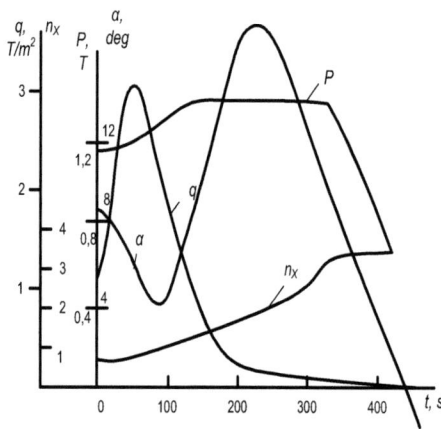

Fig 3.4. Dependencies of parameters P, α, n_x and q from time t in case of spacecraft horizontal takeoff.

For the vertical takeoff the calculations were made for the following variant of nominal initial data:

$$V_0 = 10 \, m/s, \qquad \theta_0 = 90^0, \qquad i_0 = 50.7^0, \quad \varphi_0 = 48.21^0, h_0 = 100 \, km,$$
$$m_0 = 1500t, \, n_{xadm} = 3.5, \, q_{adm} = 4000 \, kg/m^2, \, V_1 = 7.5 \, km/s, \, h_1 = 100 \, km,$$

16

$\theta_1 = 1°$, $h_{ak} = h_{\pi k} = 200$ km, $P = 2800$ t, $P_{spec} = 460.6$ s, $S_n = 0.35$ m^2.

For the horizontal takeoff, the modified initial values of velocity V_0, angle θ_0, thrust P, mass m, q_{adm} and angle θ_1 were taken as the nominal ones:

$$V_0 = 100 \text{ m/s}, \ \theta_0 = 0°, P = 1400 \text{ t}, \ m_0 = 1450 \text{ t},$$
$$q_{adm} = 3000 \text{ kg/m}^2, \ \theta_1 = 1.75°.$$

The rest of the initial data coincide with those mentioned for the vertical takeoff.

The following values of characteristic parameters were obtained in the result of calculations: for the vertical takeoff $t_v = 4.4$ s, $C = -320$ kg · deg/m^2, $\Delta\alpha = 31.5°$, $\theta^* = 19°$; for the horizontal takeoff $A = 2.25°$, $C = 5500$ kg · deg/m^2, $\Delta\alpha = 12°$, $\theta^* = 17°$.

Using the proposed control structure we determine the transfer orbit parameters h'_α and h'_π (or the corresponding values V_1, θ_1, h_1), providing relative minimum of fuel consumption for shaping the final orbit (or the relative maximum of finite mass m_f) (tables 3.1 and 3.2).

Table 3.1. Spacecraft vertical takeoff. Nominal initial data.

h'_α, km	h'_π, km	V_1, km/s	θ_1, deg	h_1, km	m_f, t
110.2	-943.8	7.5	1.00	100	205.19
143.1	-908.8	7.5	1.48	120	204.79
80.1	-981.2	7.5	0.06	80	206.79
78.0	-960.1	7.5	0.02	78	207.39
104.4	-2101.7	7.0	1.07	100	206.05
107.5	-522.6	7.65	0.85	100	205.03
78.0	-2135.8	7.0	0.03	78	207.46
200.0	-931.6	7.5	2.61	130	204.11

Table 3.2. Spacecraft horizontal takeoff. Nominal initial data.

h'_α, km	h'_π, km	V_1, km/s	θ_1, deg	h_1, km	m_f, t
130.8	-970.4	7.5	1.75	100	191.36
188.2	-953.9	7.5	2.60	120	190.31
94.0	-995.1	7.5	1.19	80	191.79
81.1	-1002.4	7.5	0.85	74	192.88
115.4	-2112.7	7.0	2.01	100	192.07
151.3	-566.4	7.65	1.72	100	191.20
74.4	-2143.2	7.0	0.34	74	192.95

200.0	-948.6	7.5	2.72	125	190.00

It was shown that for the selected set of nominal initial data for the vertical SC takeoff, it would be appropriate to insert it into the interim orbit with $h'_\alpha = 78$ km and $h'_\pi = -2135.8$ km. In this case the finite mass $m_f = 207.46$ t, which is greater by $\delta m_f = 3.35$ t than during the shaping of transfer orbit with $h'_\alpha = h_{\alpha k}$. During the horizontal takeoff the values h'_α, h'_π and m_f are, respectively 74.4 km, -2143.2 km, 192.95 t. Here the finite mass gain is $\delta m_f = 3$ t as compared with the injection scenario with altitude $h'_\alpha = h_{\alpha k}$.

In tables 3.3 and 3.4 the results are presented, which allow estimating the effect of initial data on the finite mass m_f using the proposed technique.

Table 3.3. Spacecraft vertical takeoff. Nominal initial data.

m_0, t	P, t	P_{spec}, s	n_{xadm}	q_{adm}, kg/m^2	m_f, t
1500	2800	460.6	3.5	4000	205.19
1300	2800	460.6	3.5	4000	181.30
1700	2800	460.6	3.5	4000	225.52
1500	2730	450	3.5	4000	202.36
1500	2850	470	3.5	4000	207.60
1500	2800	460.6	3.0	4000	204.78
1500	2800	460.6	4.0	4000	205.67
1500	2800	460.6	3.5	3000	203.78

Table 3.4. Spacecraft horizontal takeoff. Nominal initial data.

m_0, t	P, t	P_{spec}, s	n_{xadm}	q_{adm}, kg/m^2	m_f, t
1450	1400	460.6	3.5	3000	191.35
1300	1400	460.6	3.5	3000	177.63
1700	1400	460.6	3.5	3000	210.81
1450	1360	450	3.5	3000	185.91
1450	1430	470	3.5	3000	196.20
1450	1400	460.6	3.0	3000	190.90
1450	1400	460.6	4.0	3000	191.50
1450	1400	460.6	3.5	2000	190.04

Analyzing the results presented in the tables, we can see that mass m_f increases with increasing of start mass m_0, thrust P, specific engine thrust P_{spec}, admissible overload n_{xadm} and velocity head q_{adm}.

The analysis of received results showed the application efficiency of the proposed SC injection scenario with the preliminary shaping of transfer orbit with apogee altitude $h'_\alpha < h_{\alpha k}$ and the further two-burn SC insertion to the final orbit. The fuel mass gain as compared with the injection scenario with orbit altitude $h_{\alpha k}$ and the further apogee altitude rising to $h_{\pi k}$ for the considered variants of nominal initial made up approx. $3.0 - 3.5t$ for the vertical takeoff and approx.$2.5 - 3t$ for the horizontal takeoff.

The calculation technique of SC injection with vertical and horizontal takeoff showed its performance in the wide range of SC characteristics, engine unit, boundary conditions and limitations.

It should be noted that SC motion in the powered flight phase is carried out, as a rule, with the use of precomputed and laid in the onboard computer program of thrust vector change by the time. At the same time the developed method can be the basis for the developed adaptive board algorithms which allow prompt correcting of the control programme depending on the values of current parameters of spacecraft motion.

3. Methods of orbit determination

The functioning of large-scale orbit constellations in the earth satellite vehicle orbits requires the need in the ground control means for prompt determination and continuous adjustment of SC orbit parameters. As a result, it is necessary to search the ways for increment in efficiency of receiving the measuring data about the SC motion parameters, creation of efficient onboard calculation algorithms and reducing of operational calculations during this task solution.

Orbit determination of spacecraft to comply the specified boundary conditions is one of the major problems of flight mechanics. There are different orbit determination methods in the two-body problem [25]. The use of the mentioned methods is connected with the running of iterative processes. The approximate analytical method is proposed below, which enables to calculate the values of orbital elements by end formulae. It reduces the calculation duration by 5-8 times.

The basis of the proposed method elaboration is the transformation of end calculation dependencies received during the analytical integration of differential equations describing SC orbital motion in the velocity coordinate system, which presents a special case of the general system of equations (3.1):

$$\frac{dV}{dt} = -\frac{\mu}{r^2}\sin\theta, \qquad \frac{d\theta}{dt} = \left(\frac{V}{r} - \frac{\mu}{r^2 V}\right)\cos\theta,$$

$$\frac{dr}{dt} = V\sin\theta, \qquad \frac{d\vartheta}{dt} = \frac{V}{r}\cos\theta. \qquad (4.1)$$

here ϑ – true anomaly. The rest of symbols are similar to those in the chapter 3.

Let's turn by the formula $\frac{1}{dt} = V\sin\theta / dr$ to a new independent variable r. It should be noted that we consider spacecraft non-circular motion ($dr \neq 0$), that is, the value r can change in the range from orbital pericenter radius r_π to orbital apocenter radius r_α (when passing the pericenter or apocenter the increment dr reverses sign).

After change of variable and integration of the first equation of the system (4.1) we receive the integral of kinetic energy [26]:

$$\frac{2\mu}{r} - V^2 = C_1. \qquad (4.2)$$

We modify the second and the third equations of the system (4.1) into

$$\mathrm{tg}\,d\theta = \left(\frac{1}{r} - \frac{\mu}{r^2 V^2}\right) dr.$$

From this equation we will determine the dependence of trajectory angle θ on radius vector r:

$$\theta = \arccos \sqrt{\frac{C_2}{r(2\mu - C_1 r)}}, \tag{4.3}$$

where $C_2 = r^2 V^2 \cos^2 \theta$.

In order to obtain the law of true anomaly change ϑ of the value r we use differential equation: $d\vartheta/dr = \operatorname{ctg}\theta(r)/r$.

After its integration we have:

$$\vartheta = \vartheta_0 + \arcsin A(r) - \arcsin A(r_0), \tag{4.4}$$

where $A(r) = \mu - C_2/r/\sqrt{\mu^2 - C_1 C_2}$.

Integrating the equation $dt = dr/V(r)\sin\theta(r)$, we can find the correlation between the time of SC motion t and radius vector r:

$$t = t_0 - \frac{\mu}{C_1^{3/2}} [\arcsin B(r_1) - \arcsin B(r_0)] - C(r_1) + C(r_0), \tag{4.5}$$

where $B(r) = \dfrac{\mu - C_1 r}{\sqrt{\mu^2 - C_1 C_2}}$, $C(r) = \dfrac{\sqrt{-C_1 r^2 + 2\mu r - C_2}}{C_1}$.

The formulae (4.2) - (4.5) provide the possibility to determine the parameters and time of spacecraft motion with the known initial conditions and current values of radius vector r.

We shall use these formulae for elaboration of orbit determination method by two SC positions. Let's state the task: it is necessary to determine SC motion parameters by the known values of angular distance $\Delta\vartheta$ and flight time Δt between two points of orbit, characterized by radius vectors r_0 and r_1.

The dependencies (4.4) and (4.5) can be turned into

$$\Delta\vartheta = \arcsin\left[A(r_1)\sqrt{1 - A^2(r_0)} - A(r_0)\sqrt{1 - A^2(r_1)}\right] =$$

$$= \arcsin\left\{\frac{\sqrt{C_2}}{\mu^2 - C_1 C_2}[(\mu r_1 - C_2)C(r_0) - (\mu r_0 - C_2)C(r_1)]\right\}, \tag{4.6}$$

$$\Delta t = -C(r_1) + C(r_0) - \frac{\mu}{C_1^{3/2}}\arcsin\left[B(r_1)\sqrt{1 - B^2(r_0)} - B(r_0)\sqrt{1 - B^2(r_1)}\right] =$$

$$= -C(r_1) + C(r_0) - \frac{\mu}{C_1^{3/2}}\arcsin\left\{\frac{\sqrt{C_1}}{\mu^2 - C_1 C_2}[(\mu - C_1 r_1)C(r_0) -\right.$$

$$\left. -(\mu - C_1 r_0)C(r_1)]\right\}. \tag{4.7}$$

Let's note that the transition to the equations (4.6) and (4.7) is true for:

21

$$A(r_1)A(r_0) \geq 0 \quad \text{or} \quad A^2(r_1) + A^2(r_0) \leq 1 \qquad (4.8)$$

and

$$B(r_1)B(r_0) \geq 0 \quad \text{or} \quad B^2(r_1) + B^2(r_0) \leq 1. \qquad (4.9)$$

The analysis of expressions $A(r)$ and $B(r)$ showed that in case of increasing the radius vector r from r_π to r_α, the variable $A(r)$ changes from -1 to 1, and $B(r)$ - from 1 to -1, while $A(r) = 0$ with $r = p$ and $\Delta\vartheta = 90°$, and $B(r) = 0$ with $r = r' \leq p$ and $\Delta\vartheta = \Delta\vartheta' \leq 90°$ (the values $\Delta\vartheta'$ depending on orbit parameters are in the range from 75° to 90°). Therefore, the condition (4.8) will be surely fulfilled, if the angular distance between radius vectors r_0 and r_1 doesn't exceed 90°, and condition (4.9) – with $\Delta\vartheta \leq \Delta\vartheta' \approx 75°$.

The correlations (4.6) and (4.7) are the system of equations with two unknown quantities C_1 and C_2. One can see that it is possible to determine the integration constants C_1 and C_2 by the known values of Δt, $\Delta\vartheta$, r_0 and r_1 with the use of formulae (4.6) and (4.7) only by the iterative process.

In order to obtain the finite calculation dependencies we introduce the infinitesimal assumption of function arcsin, composing the equations (4.6) and (4.7), i.e. $\arcsin x \approx x$. In this case with $x \leq 20°$ the calculation errors do not exceed 2%, and with $x < 30°$ - they do not exceed 5%.

This assumption obviously restricts the application area of the proposed method. At the same time in rather a wide range of initial parameters Δt, $\Delta\vartheta$, r_0 and r_1, which will be detected by analyzing the numerical results, the method errors will be rather small (no more than 2-5%). Besides it will be ascertained that in this range there will always be fulfilled the conditions (4.8) and (4.9).

As a result the equation (4.6) will be as follows:

$$\Delta\vartheta = \frac{\sqrt{C_2}}{r_0 r_1 (\mu^2 - C_1 C_2)} [(\mu r_1 - C_2)C(r_0) - (\mu r_0 - C_2)C(r_1)]. \qquad (4.10)$$

After the transformation of expression (4.7) with regard to the introduced assumption we determine the dependence between the radius vectors r_0 and r_1, time slot Δt and the constants C_1 and C_2:

$$\Delta t = \frac{1}{\mu^2 - C_1 C_2} [(\mu r_1 - C_2)C(r_0) - (\mu r_0 - C_2)C(r_1)]. \qquad (4.11)$$

Dividing the equation (4.10) and (4.11), we will get rather a simple correlation between the known values of Δt, $\Delta\vartheta$, r_0 and r_1 and integration constant C_2:

$$\frac{\Delta\vartheta}{\Delta t} = \frac{\sqrt{C_2}}{r_0 r_1} \qquad \text{or} \qquad C_2 = \frac{r_0^2 r_0^2 \Delta\vartheta^2}{\Delta t^2}. \qquad (4.12)$$

22

Considering (4.12) we find the focal parameter of orbit:

$$p = C_2/\mu. \tag{4.13}$$

Solving the system of two equations

$$r_0 = \frac{p}{1 + e \cos \vartheta_0} \quad \text{and} \quad r_1 = \frac{p}{1 + e \cos(\vartheta_0 + \Delta\vartheta)},$$

we will determine the true anomaly ϑ_0, corresponding to radius vector r_0

$$\text{tg}\vartheta_0 = \text{ctg}\Delta\vartheta - r_0(p - r_1)[r_1(p - r_0) \sin \Delta\vartheta]^{-1}. \tag{4.14}$$

Knowing the values of p and ϑ_0, by the known formulae it is possible to determine other elements of desired orbit as well: eccentricity e, pericenter radius r_π and apocenter radius r_α, major semiaxis a:

$$e = (p - r_0)r_0^{-1}\cos^{-1}\vartheta_0, \quad r_\pi = p/(1 + e),$$

$$r_\alpha = p/(1 - e), \quad a = \frac{r_\pi + r_\alpha}{2} \tag{4.15}$$

Thus, by the formulae (4.12) – (4.15) it is possible to calculate elements of elliptical orbits by two positions.

Let's evaluate the calculation errors, caused by the introduction of the above-mentioned assumption and determine the range of initial conditions with which the proposed method is working with satisfactory accuracy.

The comparative analysis of numerical results of orbit elements calculation (pericenter radius r_π and apocenter radius r_α, focal parameter p, eccentricity e, major semiaxis a), received for the wide range of change in values Δt, $\Delta\vartheta$, r_0, r_1 by the proposed method and with the use of iterative method [25], showed that when Δt is changed (and, therefore $\Delta\vartheta$) during one half-turn 0.5 T the calculation errors $\delta p, \delta e, \delta a$ have clearly defined maximums, carried out with some values $\Delta t^*(0 < \Delta t^* \leq 0.5 \text{ T})$, i.e.

$$\delta p_{max} = \delta p(\Delta t^*), \quad \delta e_{max} = \delta e(\Delta t^*), \quad \delta a_{max} = \delta a(\Delta t^*).$$

At that, the values δp_{max}, δe_{max}, δa_{max} are increasing with the increase of orbit eccentricity. Thus, if for the orbit of earth satellite vehicle with the altitudes h_π=500 km and h_α=700 km ($e = 1.434 \times 10^{-2}$) maximum error values are: $\delta p_{max} = \delta a_{max} \approx 0.2\%$, $\delta e_{max} \approx 5.3\%$, then at h_π=500 km and h_α=1000 km ($e = 3.51 \times 10^{-2}$) $\delta a_{max} \approx 0.49\%$, $\delta p_{max} \approx 0.47\%$, $\delta e_{max} \approx 6.8\%$, and with h_π=500 km and h_α=5000 km ($e = 0.408$) $\delta a_{max} \approx 15.9\%$, $\delta p_{max} \approx 8.6\%$, $\delta e_{max} \approx 40\%$.

Thus, by the conducted research it is determined, that the proposed method for low-circular orbits of earth satellite vehicle ($e \leq 0.04$) has method errors, which do not exceed 2÷5% with angular distances between radius vectors r_0 and r_1, not more than 90° and not more than 0.3÷0.5% with angular distances less than 25°.

  For the orbits with $e > 0.04$ with rather big time intervals Δt, the calculation accuracy can be considerably worse. That's why it is interesting to define the maximum admissible interval Δt, with which the maximum calculation errors do not exceed a pre-established value. Considering the information, part of which is presented on figures 4.1 and 4.2, we can see that the calculation errors not exceeding 1%, can be received for the satellite orbit with parameters h_π=500 km: h_α=1000 km with $\Delta t \leq$200-300 s; h_α= 10 000 km with $\Delta t \leq$70-80 s; h_α=36 000 km with $\Delta t \leq$35-40 s.

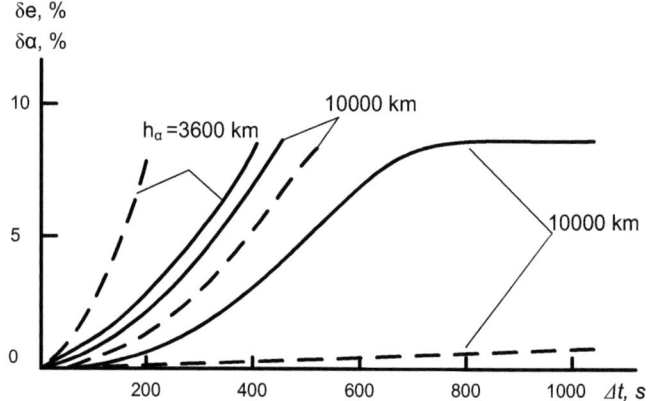

Fig. 4.1. Variation of calculation errors of eccentricity δ_e and major semiaxis δ_α in relation to spacecraft flight time between two fixed positions Δt ($h=500$ km)
Solid lines — dependencies for δ_e.
Dashed lines – dependencies for δ_α

24

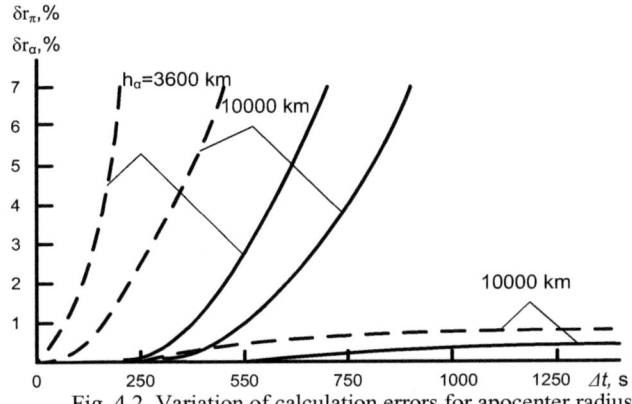

Fig. 4.2. Variation of calculation errors for apocenter radius δr_a and pericenter radius δr_π in relation to spacecraft flight time between two fixed positions Δt (h_π=500 km)

Solid lines – dependence for δr_π.

Dashed lines – dependence for δr_a

For all the calculation variants the values $A(r_1)$, $A(r_0)$, $B(r_1)$, $B(r_0)$ were determined in order to check the fulfillment of conditions (4.8) and (4.9). It has been established that in all the range of initial data with calculation errors of not more than 1÷2%, these conditions are fulfilled (table).

Calculation of parameters $A(r), B(r)$

r_0,km	r_1,km	Δt,s	$\Delta\vartheta$, deg	$A(r_0)$	$A(r_1)$	$A^2(r_0)+$ $+A^2(r_1)$	$B(r_0)$	$B(r_1)$	$B^2(r_0)+$ $+B^2(r_1)$
6921	6971	480	30	-0.48	$1.95 \cdot 10^{-2}$	0.24	-0.499	$4.7 \cdot 10^{-4}$	0.25
6921	7021	965	60	-0.48	0.516	0.5	-0.499	0.5	0.5
6921	6971	263	17	-0.78	-0.576	0.95	-0.799	-0.599	0.997
6921	7071	680	42	-0.78	-0.165	0.65	-0.799	-0.2	0.678
6921	7171	1060	65	-0.78	0.234	0.67	-0.799	0.199	0.678
6921	7271	1490	90	-0.78	0.623	1	-0.799	0.599	1
6921	6971	93	6.5	-0.96	-0.927	1.7	-0.977	-0.955	1.866
6921	7171	330	22	-0.96	-0.788	1.55	-0.977	-0.866	1.704
6921	7371	504	33	-0.96	-0.657	1.36	-0.977	-0.778	1.56
6921	8571	1280	75	-0.96	$1.86 \cdot 10^{-3}$	0.93	-0.977	-0.245	1.015
6921	6971	72	5	-0.97	-0.944	1.83	-0.983	-0.973	1.913
6921	7171	255	18	-0.97	-0.85	1.66	-0.983	-0.932	1.835
6921	7371	385	27	-0.97	-0.762	1.52	-0.983	-0.889	1.756
6921	6971	55	4.5	-0.98	-0.965	1.89	-0.997	-0.994	1.982
6921	7171	190	15	-0.98	-0.9	1.77	-0.997	-0.983	1.96

Thus, the conducted research showed high efficiency of the proposed method application for calculation of elements of keplerian orbits with low eccentricity ($e \leq$ 0.04) with angular distances between two radius vectors not exceeding $25°$. The method can be also used for elements determination of high-elliptical orbits ($e \approx$ $0.3 \div 0.7$), when SC flight time between two fixed positions is not more than $4 \div 8$ min. The calculation errors for determination of orbital parameters do not exceed 0.5%. The duration of calculations is reduced in approx. 5-8 times in comparison with the known iterative methods [25].

Collisions with space debris pose a serious threat to flight safety of operated space vehicles as well as with its small and medium-sized pieces in the space debris environment characterized by high levels of space debris density. Spacecraft control practice shows that in the existing space debris environment the medium-altitude satellites functioning at altitudes of approx. $400 \div 600$ km, $5 \div 8$ times per month approach with space debris fragments up to 15 km distance. The vehicles of $600 \div 700$ km reach the above-mentioned distance $8 \div 12$ times per month, and the vehicles of $700 \div 900$ km orbit altitudes - about $12 \div 16$ times. At that, the detection of space debris up to 10 cm by ground-based facilities is rather difficult. That is why there is a necessity of searching other ways for detection of hazardous approaches of space vehicles with different space objects.

One of the prospective methods to solve this problem is forecasting of dangerous close approaches of space vehicles with space debris using onboard facilities. Its implementation efficiency is determined by the possibility of accurate and prompt definition of space debris motion parameters using onboard optical sensors.

It is supposed that on conditions that space debris fragments are in the visibility zone of optical sensors installed on board of a SC, it is possible in time points t_i to measure distances to a piece of space debris Δr and slope angles of vector $\tilde{r}$ to the local horizon æ and SC motion plane β.

The data presented on figure 4.3 describes relative position of two objects in the SC motion plane.

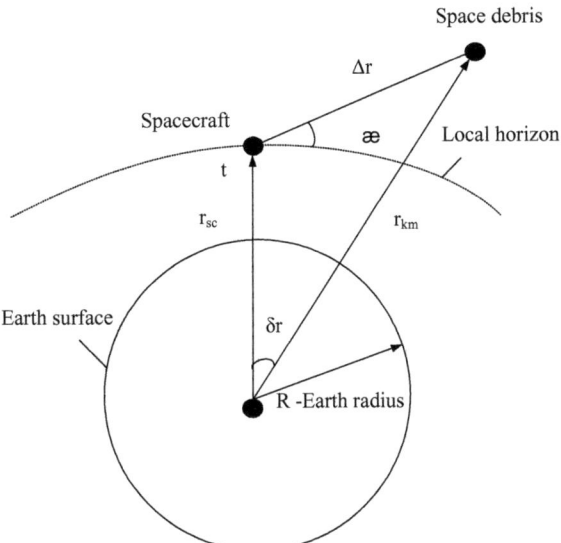

Fig. 4.3 Relative position of objects in the
spacecraft motion plane

By the known values of radius-vector of spacecraft r_{sc} and measurements of Δr and æ we calculate the radius-vector of space debris r_{sd} and angular distance δr between the current positions of SC and space debris:

$$r_{sd} = \sqrt{r_{sc}^2 + \Delta r^2 - 2 r_{sc} \Delta r \cos\left(\frac{\pi}{2} + \text{æ}\right)},$$

$$\delta r = arcsin\left[\frac{\Delta r}{r_{sd}} sin\left(\frac{\pi}{2} + \text{æ}\right)\right]$$

We calculate angular distance δr_{sc} between the vectors $r_{sc}(t_i)$ and $r_{sc}(t_{i+1})$ for two following one after another sampling instants.

Figure 4 presents the spherical triangle formed by the arcs, connecting SC positions in the moments t_i and t_{i+1} (δr_{sc}), SC and space debris positions with $t_i(\delta r)$, and also SC positions at the moment t_{i+1} and space debris positions at the moment $t_i(\sigma)$.

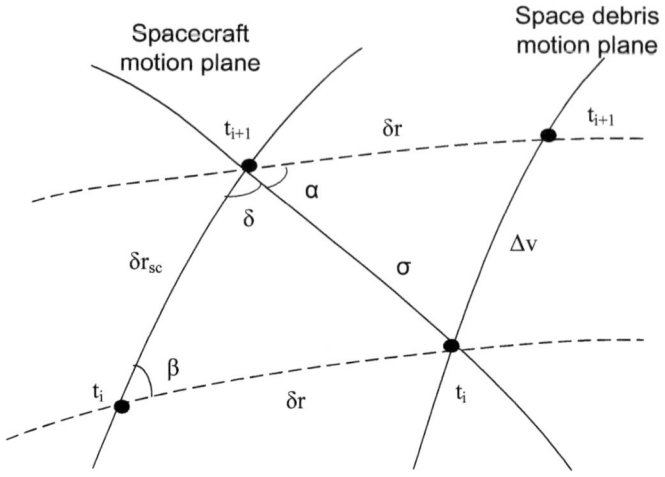

Fig. 4.4 Relative position of spacecraft motion plane and space
debris motion plane

We define the arc σ by the formulae of spherical trigonometry:
$$\sigma = arccos[\cos \delta\, r(t_i) \cos \delta\, r_{sc} - \sin \delta\, r(t_i) \sin \delta\, r_{sc} \cos \beta\, (t_i)]$$
and the angle δ between the arcs δr_{sc} and σ:
$$\delta = arcsin\left[\sin \delta\, r(t_i) \frac{\sin \beta(t_i)}{\sin \sigma}\right].$$
Then, considering the spherical triangle formed by the arc σ, and also the arcs connecting the positions of SC and space debris at the moment $t_{i+1}(\delta r)$ and space debris positions at the moments t_i and $t_{i+1}(\Delta v)$, we define the angle α between the arcs δr and σ:
$$\alpha = \pi - \beta(t_{i+1}) - \delta$$
and angular distance between the positions of space debris fragment Δv:
$$\Delta v = arccos[\cos \sigma \cos \delta\, r(t_{i+1}) - \sin \sigma \sin \delta\, r(t_{i+1}) \cos \alpha]. \qquad (4.16)$$
Thus, using the initial values of radius vectors $r_{sc}(t_i)$ and $r_{sc}(t_{i+1})$, time segment $\Delta t = t_{i+1} - t_i$ and defining by the formula (4.16) angular distances Δv, it's possible to calculate the values of focal parameter p, true anomaly ϑ_0, eccentricity e, major semiaxis a for the motion parameters of space debris according to the dependencies $(4.12 - 4.15)$.

Then we determine the inclination of space debris orbit (i). First of all we calculate the value of the course angle between the projection of SC velocity vector on the local horizon and the local parallel at the moment t_i:

28

$$\varepsilon_{sc}(t_i) = arccos\left[\frac{cos\ i_{sc}}{cos\ \varphi_{sc}(t_i)}\right].$$

Considering the measured value $\beta(t_i)$, we determine the similar course angle for the plane of conditional orbit going through the latitudes of subsatellite points of space vehicle and space debris in the moment t_i:

$$\varepsilon = \varepsilon_{sc}(t_i) - \beta(t_i).$$

We determine the inclination of this orbit (j), the arc, lying in its plane and connecting the subsatellite point of SC with the equatorial plane (δz) and flight latitude with t_i:

$$j = arccos[cos\ \varepsilon\ cos\ \varphi_{sc}(t_i)], \qquad \delta z = arcsin\left[\frac{sin\ \varphi_{sc}(t_i)}{sin\ j}\right],$$

$$\varphi_{sd}(t_i) = arcsin[sin\ j\ sin(\delta z + \delta r)].$$

On Fig. 4.5 there are the projections of spacecraft and space debris paths relative to equatorial plane.

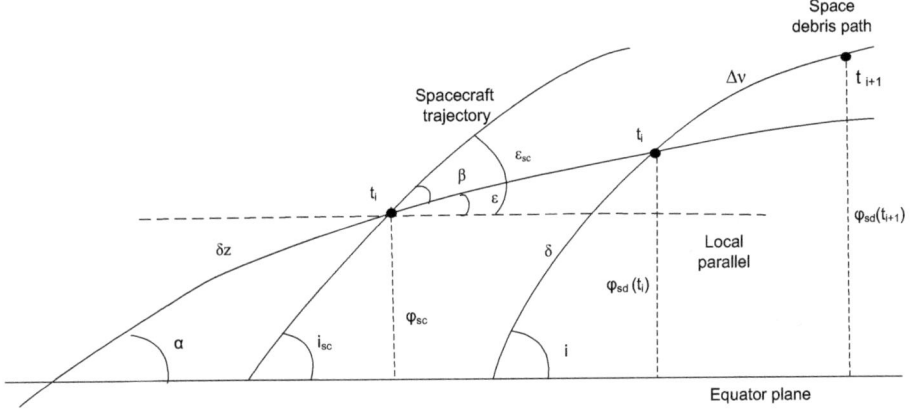

Fig. 4.5 Spacecraft and space debris paths

Then we consider two spherical triangles formed by the plane of motion of space debris, equatorial plane and also two polar planes going through the latitude $\varphi_{sc}(t_i)$ and the latitude $\varphi_{sv}(t_{i+1})$. According to the law of sines, we'll write down the following equations:

$$\frac{sin\ \varphi_{sd}(t_i)}{sin\ i} = sin\ s\ , \qquad \frac{sin\ \varphi_{sd}(t_{i+1})}{sin\ i} = sin(s + \Delta v),$$

where s – the arc lying in the motion plane of space debris and connecting its position in the moment t_i with the equatorial plane.

The joint solution of these equations allows determining the arc s and inclination of space debris orbit i:

$$s = arctg\left[\frac{sin\,\Delta v\,sin\,\varphi_{sd}(t_i)}{sin\,\varphi_{sd}(t_{i+1}) - cos\,\Delta v\,sin\,\varphi_{sd}(t_i)}\right],$$

$$i = arcsin\left[\frac{sin\,\varphi_{sd}(t_i)}{sin\,s}\right]. \qquad (4.17)$$

Thus, by the formulae (4.13-4.15, 4.17) it is possible to calculate the orbital elements of space debris by the measurement of its parameters using onboard optical sensors of SC.

The conducted research works allowed estimating the calculation errors, generated by the introduction of the above-mentioned assumptions. They also allowed defining the range of initial conditions in which the proposed method provides the required accuracy of calculations for orbit parameters of space debris. It is shown that for the low circular orbits ($e \leq 0.04$) with the measurement intervals Δt not more than approx.100 sec, the calculation errors do not exceed 0.1%. The comparable level of computational accuracy may be provided for parameter determination of high-elliptical orbits ($e \leq 0.7$), but upon condition of intervals reducing Δt to 20-25 sec. At that, considering the reiterated measurement of space debris orbit parameters and application of the known algorithms of data filtering and data smoothing, the estimated errors can be reduced.

Thus, the proposed method allows providing the orbit parameter calculation of space debris fragments in the wide variation range of initial conditions with the required calculation accuracy. At the same time, for final assessment of calculation accuracy it is necessary to take into account the perturbing factors affecting the dynamics of space objects motion: errors of performing of control inputs, measurements, changes of the ballistic coefficient.

4. Optimal SC thrust vector control in the extra-atmospheric descent phase

The control process of orbit constellations for the target application quite often requires the conducting of dynamic operations. Basing on the SC operation experience and on the research results received at the previous hierarchy levels, one can see the need for the space debris avoidance maneuvers from time to time and correction for maintaining of orbital parameters. Besides, the powered maneuvers are conducted during SC deorbiting which is the final stage for the most SC. This is the most complicated flight stage because for its successful realization it is required to conduct comprehensive research in different scientific and technical branches: ballistics, navigation, aerodynamics, heat engineering, structural integrity, etc.

At the same time, all the mentioned tasks connected with conducting of powered maneuvers are focused on the research for energy optimal schemes of SC control. This chapter is mostly devoted to optimization of SC deorbiting maneuvers and its transfer into the set area of phase space. The methodological approach to the algorithm development of this task solution may be the base for determination of optimal avoidance and adjustment maneuvers as well as for the research of optimal thrust control problem in its general formulation.

In this chapter we research the problem of SC thrust vector optimal control in the extra-atmospheric descent phase, which requires minimum fuel consumption. The two-burn control scheme of SC shows its efficiency. The fast algorithm is developed for calculation of quasioptimal descending trajectories of SC from the satellite orbit to the set area of earth surface.

As it has been noted in the previous chapters, the task solution of SC optimal control entails considerable estimate time and difficulties of computational architecture. That's why it would be reasonable to use quasioptimal algorithms of variational task solution which meet the requirements due to their effective operation speed:

The conducted research is devoted to development of such algorithms for determination of SC thrust vector optimal control in the extra-atmospheric descent phase. In particular the problem of necessary mass minimization was solved ($J = \Delta m_F = min$).

The motion of SC is described by differential equations, which is a special case of (3.1) without regard to aerodynamic forces, enabling the SC to maneuver in the atmosphere, as well as centrifugal and Coriolis forces:

$$\frac{dV}{dt} = -\frac{\rho V^2}{2P_x} - \frac{\mu}{r^2}\sin\theta + \frac{P}{m}\cos\alpha\cos\beta,$$

$$\frac{d\theta}{dt} = -\frac{\mu}{r^2 V}\cos\theta + \frac{V}{r}\cos\theta + \frac{P}{mV}\sin\alpha\cos\beta, \qquad (5.1)$$

$$\frac{d\varepsilon}{dt} = -\frac{V}{r}\cos\theta\cos\varepsilon\,tg\varphi + \frac{P}{mV\cos\theta}\sin\beta,$$

$$\frac{dr}{dt} = V\sin\theta, \qquad \frac{d\lambda}{dt} = \frac{V\cos\theta\cos\varepsilon}{r\,\cos\varphi},$$

$$\frac{d\varphi}{dt} = \frac{V}{r}\cos\theta\sin\varepsilon, \qquad \frac{dm}{dt} = -\frac{P}{P_{spec}g_E},$$

At that we used the same symbols as in the chapter 3.

Besides, in case of SC unperturbed motion with the switched off engine the correlation is true between the course angle ε and latitude φ

$$\cos\varepsilon\cos\varphi = C. \qquad (5.2)$$

The SC was controlled by change of propulsive efforts, characterized by the thrust value P and its orientation relative to SC velocity vector α and β:

$$0 \leq P \leq P_{mix}, \quad -\pi \leq \alpha \leq \pi, \quad -\pi \leq \beta \leq \pi, \qquad (5.3)$$

where α – the angle between thrust vector projection on the motion plane and SC velocity vector; β – the angle between thrust vector and SC motion plane.

The SC initial state was determined by the orbital parameters of earth satellite vehicle and its mass in the fixed moment of time t_0:

$$V_0 = V(t_0), \quad \theta_0 = 0, \quad \varepsilon_0 = \varepsilon(t_0), \quad r_0 = r(t_0), \quad \lambda_0 = \lambda(t_0),$$
$$\varphi_0 = \varphi(t_0), \quad m_0 = m(t_0). \qquad (5.4)$$

End of trajectory is the point on the earth surface ($h_R = 0$) with the specified geographic coordinates

$$\lambda_R = \lambda(h_R), \qquad \varphi_R = \varphi(h_R). \qquad (5.5)$$

The intermediate conditions were also taken into account: velocity and burnout angle at the moment when SC reaches the atmospheric conditional boundary were set ($h_{ent} = 100\ km$):

$$V_{ent} = V(h_{ent}), \qquad \theta_e = \theta(h_{ent}). \qquad (5.6)$$

By that moment the engine must finish its work. We will consider only the trajectories which end during the first half-turn. The research for the development of approximately optimal control algorithms for SC were based on the optimal control:

32

for a SC which motion is described by the equations system (5.1) and the relation (5.2), it is necessary to find the control laws $P(t)$, $\alpha(t)$, $\beta(t)$, providing extremum of functional $J = \Delta m_F = m_0 - m_f - min$ with the constraints (5.3), edge (5.4), (5.5) and intermediate (5.6) conditions.

The peculiarity of the given problem solution as compared to the SC maneuver tasks in the atmosphere is that the control process finishes before SC re-entry, i.e. the landing point deviation from the specified one is fully determined by the SC state vector at the moment of engines' closedown. It allows performing the optimal control task only in the extra-atmospheric phase, considering that the propulsion system is switched on at the initial moment of time: $P_0 = P_{max}$, the end point is determined by the conditions (5.6) and lateral displacement of the re-entry point L_{ent} in relation to the plane of initial orbit. The value L_{ent} is calculated depending on the orbital parameters (5.4) and geographic coordinates λ_R and φ_R. Then the obtained solution is united with the calculation results of deorbit coordinates and the ballistic descent trajectory simulation.

We used maximum principle of Pontryagin in order to solve the variational task [18].

We introduce the Hamiltonian

$$H = PF_1 + F_2,\qquad(5.7)$$

where

$$F_1 = \frac{\psi_1}{m}\cos\alpha\cos\beta + \frac{\psi_2}{mV}\sin\alpha\cos\beta + \frac{\psi_3}{mV\cos\theta}\sin\beta - \frac{\psi_7}{P_{spec}g_E}$$

$$F_2 = -\frac{\mu}{r^2}\sin\theta\psi_1 - \frac{\mu}{r^2V}\cos\theta\psi_2 + \frac{V}{r}\cos\theta\psi_2 - \frac{V}{r}\cos\theta\cos\varepsilon\,tg\varphi\psi_3 +$$

$$+V\sin\theta\psi_4 + \frac{V\cos\theta\cos\varepsilon}{r}\frac{}{\cos\varphi}\psi_5 + \frac{V}{r}\cos\theta\sin\varepsilon\,\psi_6.$$

The conjugate variables ψ_i $(i = 1,2,\dots,7)$ are determined by the following relations:

$$\frac{d\psi_1}{dt} = -\frac{\partial H}{\partial V} = \frac{\rho V\psi_1}{P_x} - \frac{\mu}{r^2V^2}\cos\theta\,\psi_2 - \frac{1}{r}\cos\theta\psi_2 + \frac{P}{mV^2}\cdot$$

$$\cdot\sin\alpha\cos\beta\,\psi_2 + \frac{1}{r}\cos\theta\cos\varepsilon\,tg\varphi\psi_3 +$$

$$+\frac{P}{mV^2\cos\theta}\sin\beta\,\psi_3 - \sin\theta\,\psi_4 - \frac{1}{r}\frac{\cos\theta\cos\varepsilon}{\cos\varphi}\psi_5 - \frac{1}{r}\cos\theta\sin\varepsilon\psi_6,$$

$$\frac{d\psi_2}{dt} = -\frac{\partial H}{\partial\theta} = \frac{\mu}{r^2}\cos\theta\psi_1 - \frac{\mu}{r^2V}\sin\theta\psi_2 + \frac{V}{r}\sin\theta\psi_2 -$$

33

$$-\frac{V}{r}\sin\theta\cos\varepsilon\,tg\varphi\psi_3 - \frac{P}{mV\cos^2\theta}\sin\theta\sin\beta\psi_3 -$$
$$-V\cos\theta\,\psi_4 + \frac{V\sin\theta\cos\varepsilon}{r}\frac{1}{\cos\varphi}\psi_5 + \frac{V}{r}\sin\theta\sin\varepsilon\,\psi_6, \qquad (5.8)$$

$$\frac{d\psi_3}{dt} = -\frac{\partial H}{\partial\varepsilon} = -\frac{V}{r}\cos\theta\sin\varepsilon\,tg\varphi\psi_3 + \frac{V\cos\theta\sin\varepsilon}{r}\frac{1}{\cos\varphi}\psi_5 - \frac{V}{r}\cos\theta\cos\varepsilon\,\psi_6,$$

$$\frac{d\psi_4}{dt} = -\frac{\partial H}{\partial r} = -\frac{2\mu}{r^3}\sin\theta\,\psi_1 - \frac{2\mu}{r^3 V}\cos\theta\,\psi_2 + \frac{V}{r^2}\cos\theta\psi_2 - \frac{V}{r^2}\cdot$$
$$\cdot\cos\theta\cos\varepsilon tg\varphi\,\psi_3 + \frac{V\cos\theta\cos\varepsilon}{r^2}\frac{1}{\cos\varphi}\psi_5 + \frac{V}{r^2}\cos\theta\sin\varepsilon\psi_6.$$

$$\frac{d\psi_5}{dt} = -\frac{\partial H}{\partial\lambda} = 0 \qquad \frac{d\psi_5}{dt} = -\frac{\partial H}{\partial\varphi} = \frac{V\cos\theta\cos\varepsilon}{r}\frac{1}{\cos^2\varphi}\psi_3 - \frac{V\cos\theta\cos\varepsilon\sin\varphi}{r}\frac{1}{\cos^2\varphi}\psi_5,$$

$$\frac{d\psi_7}{dt} = -\frac{\partial H}{\partial m} = \frac{P}{m^2}\cos\alpha\cos\beta\psi_1 + \frac{P}{m^2 V}\sin\alpha\cos\beta\,\psi_2 + \frac{P}{m^2 V\cos\theta}\sin\beta\,\psi_3 .$$

The laws of parameter variation α, β and P with optimal control are determined from the Hamiltonian maximization condition. The relations for calculation of optimal values α and β are obtained from the conditions $\partial H/\partial\alpha = 0$ and $\partial H/\partial\beta = 0$:

$$tg\alpha = \frac{\psi_2}{V\psi_1}, \qquad (5.9)$$

$$tg\beta = \frac{\psi_3}{(V\psi_1\cos\alpha + \psi_2\sin\alpha)\cos\theta} = \frac{\psi_3\cos\alpha}{V\psi_1\cos\theta} . \qquad (5.10)$$

With a help of inequation $\partial^2 H/\partial\alpha^2 < 0$, $\partial^2 H/\partial\beta^2 < 0$ we will establish the membership of angles α and β to one of the two quadrants:

$$\cos\alpha = sign\,\psi_1 , \qquad (5.11)$$

$$\cos\beta = sign\left(\psi_1\cos\alpha + \frac{\psi_2}{V}\sin\alpha\right) = sign\left(\frac{\psi_1}{\cos\alpha}\right) . \qquad (5.12)$$

The engine thrust possesses the boundary values:

$$\left.\begin{array}{l} P = P_{max} \;\; \text{with } F_1 > 0 \\ P = 0 \quad\;\; \text{with } F_1 < 0 \end{array}\right\} \qquad (5.13)$$

Let's prove that there are no more than two powered flight phases.

The expression F_1 has a switchover function in the thrust optimal control. In order to determine the number of burns it is necessary to examine the function F_1, which, according to [27], is calculated from the equation

$$F_1 = \frac{1}{m}\left(\psi_1\cos\alpha\cos\beta + \frac{\psi_2}{V}\sin\alpha\cos\beta + \frac{\psi_3}{V\cos\theta}\sin\beta\right).$$

Considering the formulae (5.9) and (5.10) we get

$$F_1 = \frac{1}{m}\left(\frac{\psi_1}{\cos\alpha\cos\beta}\right). \qquad (5.14)$$

In order to investigate the behavior of curve F_1 we transform the mathematical model of SC motion, as the equation (5.14) cannot be solved analytically. Let's consider that SC flight with the running engine is determined only by active forces and during the coast flight – only by gravitational forces. Suppose that angles α and β, as well as trajectory angle θ change insignificantly during the powered flight. Then the differential equations for conjugate variable ψ_1, influencing the function F_1, take the form:

$$\frac{d\psi_1}{dt} = \frac{P}{mV^2}\sin\alpha\cos\beta\,\psi_2 + \frac{P}{mV^2}\sin\beta\,\psi_3 \quad \text{with } P \neq 0,$$

$$\frac{d\psi_1}{dt} = -\frac{\mu}{r^2V^2}\psi_2 - \frac{1}{r}\psi_2 \quad \text{with } P = 0.$$

In contexts of the made assumptions in both cases $\dot{\psi}_2 = 0$, that is $\psi_2 = C^*$.

We show that $\dot{\psi}_1(t) \geq 0$ with $t_0 \leq t \leq t_R$. The equation for ψ_1 with $P \neq 0$ considering (5.9) and (5.10) is transformed in the following way:

$$\frac{d\psi_1}{dt} = \frac{P\psi_1}{mV\cos\alpha\cos\beta}(\sin^2\alpha\cos^2\beta + \sin^2\beta).$$

As the expression $\psi_1/\cos\alpha$ has the same sign as $\cos\beta$ (q.v. (5.12)), the inequality is true $\dot{\psi}_1 \geq 0$ with $P \neq 0$.

The sign of the variable ψ_1 with $P = 0$ depends on the sign of the constant $\psi_2 = C^*$, which is determined from the following considerations. In order to transfer SC from earth satellite orbit to a descending trajectory, angle α in the process of powered flight should be in the range of $-\pi \leq \alpha < -\pi/2$. Then from the equation (5.11) we obtain that $\psi_1 \leq 0$, and with the help of equation (5.9) determine that $\psi_2 = C^* \leq 0$. Hence, $\dot{\psi}_1 \geq 0$ during the coast flight. At that, the analysis of dependence $\dot{\psi}_1$ showed that the conjugate variable ψ_1 jumps at the moment of engine thrust P switching. Hence we can conclude that the variable ψ_1 during all the considered flight phase of SC changes its sign no more than two times.

Thus, considering the condition (5.14) we may conclude that the maximum number of zeros of function F_1, as well as the number of SC engine thrust switchings, equals two. Besides in this case the switchings are made from $P = P_{max}$ to $P = 0$, and then to $P = P_{max}$ again.

Thus, the laws of thrust vector optimal control are characterized by the dependencies (5.9), (5.10), (5.13). In order to obtain the numerical evaluation of optimal trajectory it's necessary to solve the boundary value problem which consists in iteration of 10 ($N=10$) parameters at the beginning of a trajectory (seven-parameters vector of conjugate variables ψ_{i0} and values of control functions P_0, α_0, β_0), with which the final edge conditions are fulfilled.

Some boundary values of conjugate variables can be obtained on account of transversability condition [27]:

$$[(\psi_1 - 1)\delta m - H\delta t + \psi_3\delta\varepsilon + \psi_5\delta\lambda + \psi_6\delta\varphi]_{t_0}^{t_k} = 0. \tag{5.15}$$

The rest of summands cannot be determined due to the fact that the corresponding variables are fixed and their variations at the trajectory end points are equal to zero.

As the variations $\delta\varepsilon$ and $\delta\varphi$ with $t = t_0$ are interdependent ones, the transversality condition is fulfilled in case if the following equality occurs

$$\psi_{30}\delta\varepsilon + \psi_{60}\delta\varphi = 0. \tag{5.16}$$

On the other hand, the variations $\delta\varepsilon$ and $\delta\varphi$ are related by

$$\frac{\partial q}{\partial\varepsilon}\delta\varepsilon + \frac{\partial q}{\partial\varphi}\delta\varphi = 0$$

where $q = \cos\varepsilon\cos\varphi - C$.

Therefore the condition (5.16) will be achieved if the initial values ψ_{30} and ψ_{60} are chosen to realize the relation

$$\left(\frac{\partial q}{\partial\varphi}\right)_0 \psi_{30} = \left(\frac{\partial q}{\partial\varepsilon}\right)_0 \psi_{60} \qquad \text{or} \qquad \psi_{60} = tg\varphi_0 ctg\varepsilon_0\psi_{30}. \tag{5.17}$$

With $t = t_R$ the equation (5.15) due to arbitrariness of variations δm, δt, $\delta\lambda$ is possible only if

$$\psi_{7R} = 1, \quad H = 0, \quad \psi_{5R} = 0.$$

As the variables t and λ do not constitute in an explicit form the right parts of the system (5.1), we can conclude that, the Hamiltonian H and the conjugate variable ψ_5 are identically equal to zero:

$$H \equiv 0, \quad \psi_5 \equiv 0. \tag{5.18}$$

The equations (5.9), (5.10), (5.17), (5.18) and the equality defined in the problem statement $P(t_0) = P_{max}$ provide six constraint equations between ten initial parameters. The deficient four equations necessary for initial estimate of boundary problem solution can be got introducing the assumptions about vanishing of switching function at the initial instant and impulse character of engines' operation. From the first assumption follows:

36

$$\frac{\psi_{10}}{m_0}\cos\alpha_0\cos\beta_0 + \frac{\psi_{20}}{m_0V_0}\sin\alpha_0\cos\beta_0 + \frac{\psi_{30}}{m_0V_0\cos\theta_0}\sin\beta_0 - \frac{\psi_{70}}{P_{spec}g_E} = 0. \quad (5.19)$$

In view of equation (5.18) and (5.19) we get:

$$-\frac{\mu}{r_0^2}\sin\theta_0\psi_{10} - \frac{\mu}{r_0^2V_0}\cos\theta_0\,\psi_{20} + \frac{V_0}{r_0}\cos\theta_0\psi_{20} - \frac{V_0}{r_0}\cos\theta_0\cos\varepsilon_0 tg\varphi_0\psi_{30} +$$

$$+V_0\sin\theta_0\,\psi_{40} + \frac{V_0\cos\theta_0\cos\varepsilon_0}{r_0}\frac{1}{\cos\varphi_0}\psi_{50} + \frac{V_0}{r_0}\cos\theta_0\sin\varepsilon_0\psi_{60} = 0. \quad .20)$$

Using the second equation and solving the Keplerian motion equations [26], we may determine initial orientation angles of thrust vector α_0 and β_0 at the moment of SC reentry:

$$\alpha_0 = -\pi + \arcsin\left(V_1\sin\frac{\theta_1}{\Delta V}\right). \quad (5.21)$$

where

$$V_0 = \sqrt{\frac{\mu}{r_0}}, \qquad V_1 = \sqrt{V_{ent}^2 + \frac{2\mu}{r_0} - \frac{2\mu}{r_{ent}}}, \qquad r_0 = r_1 = R_E + h_0,$$

$$\cos\theta_1 = \frac{r_{ent}V_{ent}\cos\theta_{ent}}{r_1V_1}, \qquad \Delta V = \sqrt{V_0^2 + V_1^2 - 2V_0V_1\cos\theta_1}, r_{ent} = R_E + h_{ent},$$

$$\beta_0 = arctg\left\{\frac{L_{ent}}{R_E\left[\arcsin\left(\frac{\mu - C_2/r_{ent}}{\sqrt{\mu^2 - C_1C_2}}\right) - \arcsin\left(\frac{\mu - C_0/r_0}{\sqrt{\mu^2 - C_1C_2}}\right)\right]}\right\}, \quad (5.22)$$

where

$$C_1 = \frac{2\mu}{r_{ent}} - V_{ent}^2, \qquad C_2 = r_{ent}^2 V_{ent}^2\cos^2\theta_{ent}.$$

If the single-burn SC transfer to the finite point with the coordinates $r = r_{ent}$, $V = V_{ent}$, $\theta = \theta_{ent}$ and $L_L = L_{ent}$ (lateral entering) is impossible, the two-burn transfer is considered. The first burn of the value ΔV_1 with $\alpha_0 = \pi$ provides SC reentry with the specified values V_{ent} and L_{ent}, and with the help of the second burn ΔV_2, given $r = r_{ent}$, the final value θ is adjusted. At that the initial value of angle β_0 is always determined by the formula (5.22).

Thus, the mentioned analytical dependencies enable us to find the first approximation for the boundary task solution.

The conducted analysis of numerical results, part of which is presented on fig. 5.1 and 5.2, reveals the structure of thrust vector optimal control with minimum of active mass. The calculations were made with variation of altitude h_0 and inclinations i_0 of earth satellite circular orbits, initial SC mass m_0, reduced frontal surface load P_x, thrust P and specific thrust P_{spec} of the propulsion device, spacecraft reentry conditions V_{ent}, θ_{ent}, L_{ent} in the range:

$$300 \leq h_0 \leq 700 \text{ km,} \qquad 40^0 \leq i_0 \leq 80^0,$$
$$500 \leq m_0 \leq 2500 \text{ kg,} \qquad 500 \leq P_x \leq 200 \ kg/m^2,$$
$$1000 \leq P \leq 5000 \text{ kg,} \qquad 250 \leq P_{spec} \leq 450 \text{ s,}$$
$$6{,}5 \leq V_{ent} \leq 8 \text{ km/s,} \quad -15^0 \leq \theta_{ent} \leq -5^0, \qquad 0 \leq L_{ent} \leq 800 \text{ km.} \qquad (5.23)$$

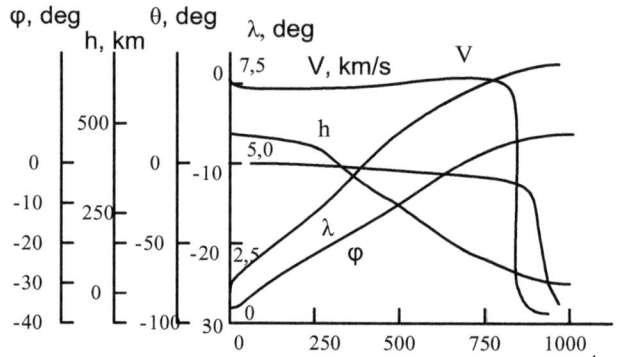

Fig. 5.1. Change of phase coordinates
from deorbiting time

$$h_0 = 500 \text{ km,} \ i_0 = 55^0, \quad m_0 = 2000 \text{ kg,} \quad P_x = 1000 \text{ kg/m}^2, \quad P = 2000\text{kg,}$$
$$P_{spec} = 320 \text{ s,} \qquad V_{ent} = 7{,}4 \text{ km/s,} \qquad \theta_{ent} = -8^0$$

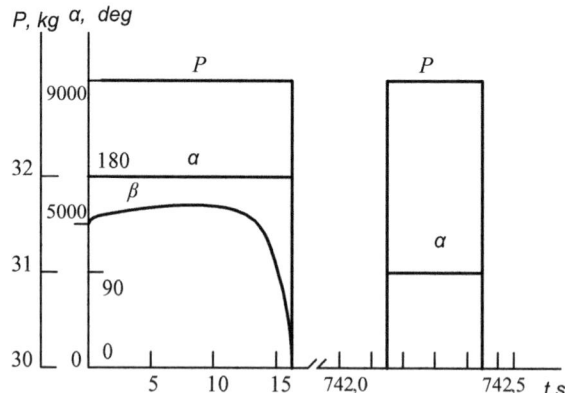

Fig. 5.2. Change of control parameters P, α, β from
deorbiting time (the same variant of initial data)

38

The following values were used as nominal values of varied parameters:
$$h_0 = 500 \text{ km}, i_0 = 55^0, \quad m_0 = 2000 \text{ kg}, \quad P_x = 1000 \text{ kg/m}^2,$$
$$P = 2000 \text{ kg}, \quad P_{spec} = 320 \text{ s}, \quad V_{ent} = 7.4 \text{ km/s}, \quad \theta_{ent} = -11^0. \quad (5.24)$$
It is showed that for all the variation range of parameters the optimal control consists in two- burn ignition: at the 1st burn the SC is transferred from the satellite orbit to the descending trajectory, at the 2nd burn the reentry parameters are corrected. The thrust vector orientation at the 1st burn consists in the following: angle α is constant and equals approx. 180^0, angle β slightly changes from the initial value β_0, lying in the range from 120^0 to 180^0, depending on the value of lateral displacement L_{ent}, by $\pm 1 \div 2^0$.

The analysis of results provides opportunity to establish general principles of optimal control and derive a non-iterative algorithm on their basis. It involves the use of control programs with the constant thrust vector orientation at 1st burn: $\beta = \beta_0$, where β_0 is calculated by the formula (5.23).

The finish of 1st burn corresponds to SC velocity to the value providing during the further coasting flight the reentry with the specified velocity V_{ent}. At the 2nd burn, angle $\beta \approx 0$, and angle α either equals 90^0 if there's a need to reduce angle $|\theta_{ent}|$, or $\alpha = -90^0$ with the increase of $|\theta_{ent}|$. The start of 2nd burn is chosen so that its finish will correspond to the moment of SC reentry ($h = h_{ent} = 100$ km). As one would expect, the duration of the 1st burn is considerably greater than of the 2nd burn.

Application of such algorithm will not lead to the considerable increase in fuel consumption Δm_F in comparison with Δm_{Fmin}: the differences do not exceed $1 \div 2\%$. The supposed control will be called approximately optimal control.

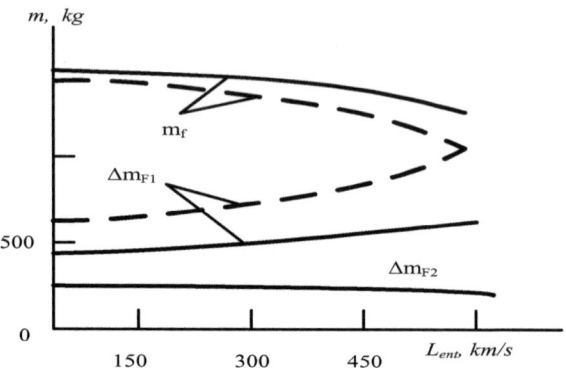

Fig. 5.3. Dependencies of SC final mass and mass Δm_{F1}
and Δm_{F2} on the lateral displacement L_{ent} (the variant of
nominal initial data)
Solid lines – approximately – optimal control;
dashed lines – the one-burn scheme

Figure 5.3 presents the dependencies of SC finite mass m_f, spent at the 1st and 2nd burns m_{F1} and Δm_{F2}, on the lateral displacement L_{ent}. For comparison there are the results of fuel mass calculation Δm_{F1}, necessary with the use of spacecraft one-burn deorbiting scheme and corresponding to the spacecraft finite mass m_f. Analyzing this data, we should note a considerable efficiency of the proposed deorbiting scheme. Thus, the increase of SC finite mass δm_f is approx. 180 kg.

As we should expect, with approximately optimal control the fuel mass Δm_{F1} monotonically increases with the increase of value L_{ent}: the change of L_{ent} from 0 to 600 km leads to increase of mass Δm_{F1} from 430 to 660 kg. The fuel mass Δm_{F2} changes less: in the same variety range L_{ent} mass Δm_{F2} reduces from 190 to 160 kg. On the whole for nominal values of variable parameters (5.24) the SC finite mass is approx. 1200-1400 kg.

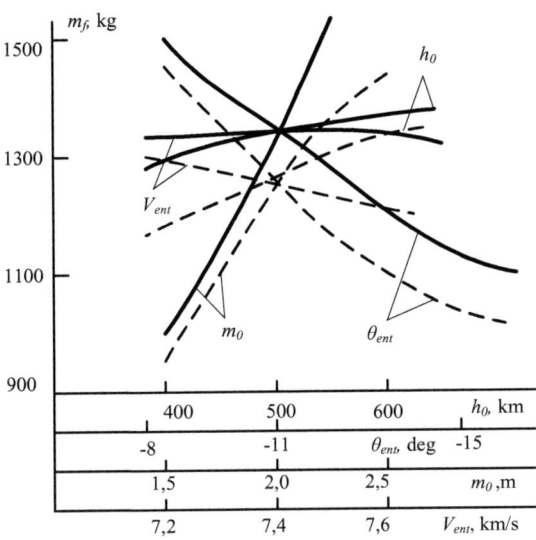

Fig. 5.4. Dependencies of SC final mass on altitude h_0, initial mass
m_0, velocity V_{ent}, trajectory angle θ_{ent}
Solid lines – approximately – optimal control;
dashed lines – the one-burn scheme

The results presented on Fig. 5.4 demonstrate specific influence of variable parameters $\left(h_0, m_0, V_{ent}, \theta_{ent}\right)$ on the finite mass m_f and on the efficiency of mass increase δm_f using two-burn SC deorbiting pattern. Considering them one can see that the finite mass m_f is increasing with the altitude increase of earth satellite orbit h_0, initial mass m_0 and with reducing of SC reentry velocity V_{ent} and absolute value of trajectory angle θ_{ent}. Thus, the change of h_0 from 400 to 600 km leads to increase of m_f from 1265 to 1380 kg, mass change m_0 from 1.5 to 2.3 t leads to increase of m_f from 995 to 1660 kg, reducing of reentry velocity V_{ent} from 7.6 to 7.2 km/s leads to increase of m_f from 1310 to 1325 kg, change of trajectory angle from -15^0 to -8.5^0 – results in the increase of m_f from 1110 to 1490 kg (L_{ent}=300 km).

The power characteristics of engine P and P_{spec}, reduced frontal surface load P_x and inclination of earth satellite orbit i_0 scarcely affect the finite mass of SC.

The efficiency of δm_f increase due to application of the proposed scheme is provided in all the variation range of variable parameters, presented on fig. 5.4.

Besides the high intensity of m_f increase is revealed for greater values m_0, absolute values θ_{ent} and smaller altitutudes h_0: δm_f achieves approx. 270 kg.

Thus the represented materials show the possibility and big power gain of SC two-burn deorbiting pattern in the wide range of boundary conditions, design, mass and power characteristics of SC and engine. It should be noted that the proposed methodological approach can be applied also for the solution of tasks of thrust vector optimal control during interorbital maneuvers and correction for maintenance of SC orbital parameters in the specified limits.

Consider the possibility of deriving fast algorithms for computing the reentry SC ballistic descent trajectory from its orbit to a predetermined earth surface landing point using the proposed SC thrust vector control pattern. It is known that the landing point parameters of free-flight vehicles depend on the SC atmosphere entry phase parameters. That is why the problem of SC landing in the predetermined earth surface landing point reduces to the determination of SC deorbiting coordinates $\lambda_{deorbit}$ and $\varphi_{deorbit}$ and relevant lateral displacement of the descent trajectory L_{ent} at atmosphere entry relative to the orbit plane on the assumption of perturbing factors lack. We shall describe the methodology of determination of values $\lambda_{deorbit}$, $\varphi_{deorbit}$ and L_{ent}.

Suppose that the boundary conditions (5.4) and (5.5) and input data (5.24) have the values at which the spacecraft may deorbit and descent to a predetermined earth surface landing point from the current orbit, i.e. the landing point lateral displacement optimally to the SC orbit plane L_{lf} (lateral finite) does not exceed the maximum realizable L_{lmax} (in other cases the descent orbit is to be determined).

Find the L_{lf} functional dependence on the boundary conditions and input data. Suppose that at a certain time t_{Eq} the spacecraft is passing the equator and is characterized by the subsatellite point with $\lambda = \lambda_{Eq}$, $\varphi = 0$ geographic coordinates. The terminal point B coordinates are φ_B and λ_B (Fig. 5.5).

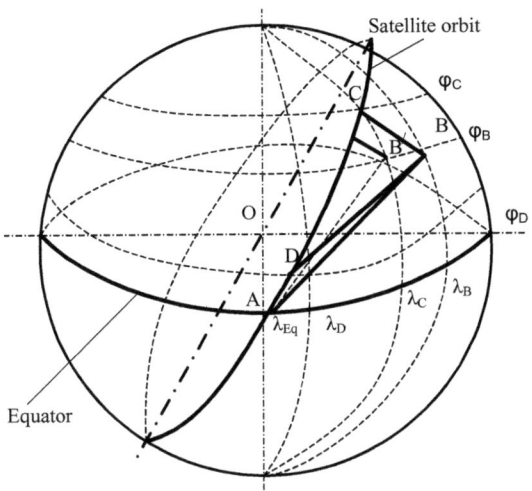

Fig. 5.5. Pattern for SC deorbiting coordinates determination

A – SC equatorial position, B – SC landing point, C – cross point of AOC and BOC orthogonal planes, D – deorbiting point, B' - point B drift due to Earth rotation for time t_1.

First determine the angular distance between points A and B for non-rotating Earth applying the known spherical trigonometry formula:

$$\Delta v = \arccos(\cos \varphi_B \cos \Delta\lambda), \qquad (5.25)$$

where $\Delta\lambda = \lambda_B - \lambda_A$.

The specified Δv value may be found by formula (5.25), introducing in it $\Delta\lambda$ obtained taking into account the Earth rotation (Fig. 5.5):

$$\Delta\lambda = \lambda_B + \omega_E t_1 - \lambda_A,$$

where

$$t_1 = \Delta v r_0^{3/2} \sqrt{\mu}. \qquad (5.26)$$

Resorting to the analytical recurrent relations (5.25) and (5.26) two or three times one may obtain an angular distance with errors not exceeding approx. 1% with negligent computation burden.

Then write the equation for computing the conditional orbit inclination passing through points A and B

$$i_{spec} = \arcsin\left(\frac{\sin \varphi_B}{\sin \varphi \Delta v}\right).$$

43

Applying the theorem of sines for the spherical triangle ABC determine the angular distance b between B and C:

$b = \arcsin(\sin \Delta v \sin \Delta i)$, where $\Delta i = i_0 - i_{spec}$

It is obvious that there is a dependence of the sought value L_{lf} and angular distance b

$$L_{lf} = bR_E. \qquad (5.27)$$

First calculate the longitudinal descent range L_R composed of the SC extra-atmospheric flight leg L_{ext} and atmospheric flight leg L_{atm} for determining the SC deorbiting point coordinates with which the needed point L_{lf} lateral displacement is provided for the identified control structure. The L_{ext} value is calculated in the impulse problem statement applying the formula

$$L_{ext} = R_E\left[\arcsin\left(\frac{\mu - C_2/r_0}{\sqrt{\mu^2 - C_1 C_2}}\right)\right].$$

$$C_1 = \frac{2\mu}{r_{ent}} - V_{ent}^2, \qquad C_2 = r_{ent}^2 V_{ent}^2 \cos^2\theta_B.$$

The free-flight space vehicle atmospheric flight range may be computed applying the system numerical integration (5.1) with $P=0$ for the set atmosphere entry conditions. However the following approximated pattern is proposed in order to reduce the computation time. Note that for the SC atmospheric flight the motion equation system for free-flight SC differs from the Kepler motion differential equations by presence of the first sum in the velocity equation characterizing the aerodynamic forces action $-\rho V^2/2P_x$. The pattern substance lies in that the analytical solution of Kepler equations is considered as a reference solution shown, in particular, in [26]:

$$V^{(R)} = \sqrt{\frac{2\mu}{r} - C_1}, \qquad \theta^{(R)} = \arccos\sqrt{\frac{C_2}{r(2\mu - C_1 r)}}, \qquad (5.28)$$

$$L^{(R)} = R_E\left[\arcsin\left(\frac{\mu - C_2/r}{\sqrt{\mu^2 - C_1 C_2}}\right) - \arcsin\left(\frac{\mu - C_2/r_{ent}}{\sqrt{\mu^2 - C_1 C_2}}\right)\right]$$

The current velocity values $V^{(R)}$, trajectory angle $\theta^{(R)}$, atmosphere leg range $L^{(R)}$ are computed using these formulae for any height values h from $h_{ent} = 100$ km to $h_R = 0$.

Only the velocity in its explicit form out of all SC flight phase coordinates is a function of the aerodynamic forces (the remaining coordinates are under indirect

44

effect of the aerodynamic forces). That's why we shall compute the trajectories using the formulae (5.28) for the argument variation intervals h, where the aerodynamic forces are assumed to be constant preliminarily recounting the flight velocity and corresponding constants C_1 and C_2 values.

The dependence for the flight velocity and with due regard for the aerodynamic forces effect will be written as follows:

$$V = \sqrt{\frac{2\mu}{r} - C_1 - \int_{t_{ent}}^{t} \frac{\rho V^2}{2P_x} dt}. \tag{5.29}$$

Using the assumption concerning the exponential character of the atmosphere density variations dependence on the height

$$\rho = \rho(h = 0)\exp(-\beta h).$$

the third equation of the system (5.1) is transformed in the following way:

$$\frac{d\rho}{dt} = -\rho\beta V \sin\theta.$$

After substitution of the variable $dt = -\frac{d\rho}{\rho\beta V}\sin\theta$ and mentioned introduction of the piecewise constancy intervals the formula (5.29) will be as follows:

$$V = \sqrt{\frac{2\mu}{r} - C_1 + \frac{V(\rho - \rho_{ent})}{2P_x\beta\sin\theta}}.$$

Solving the equation in relation to the variable V, write the ultimate dependence for computing the free-flying space vehicle atmospheric flight:

$$V = \sqrt{\frac{2\mu}{r} - C_1 \left(1 - \frac{\rho - \rho_{ent}}{2P_x\beta\sin\theta}\right)^{-1}}. \tag{5.30}$$

We turn our attention to argument variation interval Δh selection at which the assumption concerning the aerodynamic forces piecewise constancy remains in force. When selecting the value Δh one should take into consideration the requirements to the admissible computation errors and computation fastness. The presented numerical computation analysis shows that for the variable parameters range (5.23) and $\Delta h = 10$ km the computation errors of ballistic descent δL do not exceed approx.1%. In this case the computation length reduces more than by an order of magnitude as compared with the numerical computation methods.

So, after identification of the terminal point lateral displacement relative to the orbit plane L_{lf} and longitudinal descent range L_R, consisting of the sum of extra-atmospheric and atmospheric flight legs L_{ext} and L_{atm} change over to determination of the SC deorbiting coordinates $\lambda_{deorbit}$ and $\varphi_{deorbit}$ depicted in Fig. 5.5 by benchmark D. Considering the spherical triangle DCB we obtain the formula for computing the deorbiting coordinates $\lambda_{deorbit}$ and $\varphi_{deorbit}$:

$$\varphi_{deorbit} = \varphi_D = \arcsin[\sin i_0 \sin(l - d)],$$
$$l = \arccos[\cos \Delta v / \cos(L_{lf}/R_E)], \qquad (5.31)$$
$$d = \arccos[\cos(L_R/R_E)/\cos(L_{lf}/R_E)],$$
$$\lambda_{deorbit} = \lambda_D = \lambda_{Eq} + \arccos[\cos(l - d)/\cos\varphi_D].$$

Then find the lateral displacement of SC atmosphere entry point relative to the orbit plane L_{ent}, which being the input parameter for SC trajectory computation on its extra-atmospheric flight leg will ensure the required displacement L_{lf} on the earth surface:

$$L_{ent} = R_E \arcsin[\sin(L_{ext}/R_E) \sin \Delta\varepsilon],$$

where

$$\Delta\varepsilon = \arccos(\cos i_0 / \cos \varphi_D) - \arccos(\cos i^* / \cos \varphi_D),$$

$$i^* = \arcsin\left\{ \frac{\sin \varphi_D}{\sin\left[arctg\left(\frac{\sin(L_f/R_E) \sin \varphi_D}{\sin \varphi_B - \cos(L_f/R_E) \sin \varphi_D} \right) \right]} \right\}. \qquad (5.32)$$

Using the analytical formulae (5.29)-(5.32) and input data and conditions we calculate the SC deorbiting coordinates and required lateral displacement L_{ent}, which together with application of the thrust vector control on the extra-atmospheric flight leg ensures the descent of ballistic type SC to a certain earth surface area near the point with predetermined coordinates φ_R and λ_R (5.5).

The presented materials make it possible to reduce the stated optimal control problem to a noniterative problem of equations modeling (5.1). In this case the parameters unequivocally determining the SC control structure are computed using analytical formulae. Due to introduced assumptions for these formulae derivation it would be practicable to assess possible deviations of SC landing points from the required point. The studies made within a wide range of variable parameters change (5.23) show that the SC landing point deviations lie within $2 \div 3$ km range on the average, sometimes reaching approx. 5 km. To diminish these deviations to values below 1 km computations are made for trajectory specification when corrections of longitudinal and lateral point deviations δL and δL_l accordingly computed during the computation run relative to the predetermined point with $= \varphi_R$ and $\lambda = \lambda_R$ are introduced in the formulae for calculating SC deorbiting coordinates (5.31) and lateral SC atmosphere entry point displacement (5.27), (5.32).

The performed studies of optimal thrust vector control assuring a free-flight vehicle deorbiting and landing in a predetermined earth surface area with a maximal propellant consumption enable to make the following conclusions:

– the optimal SC thrust vector control structure to be used on the extra-atmospheric deorbiting leg with a minimal propellant consumption was determined as a result of solving the variational problem;

– high efficiency of the SC two-impulse control pattern on its extra-atmospheric flight leg has been demonstrated. The saving of propellant mass needed for SC transfer from its NES orbit to atmosphere entry trajectory with the set velocity, trajectory angle and lateral displacement values as compared with the single-impulse descent pattern for the variable parameters range (5.23) in certain cases reaches $200 \div 270$ kg;

– the fast algorithm for computing approximated/optimal trajectories of spacecraft deorbiting and descending to a predetermined earth surface area has been derived. The duration of noniterative computation of landing point deviations from the predetermined point is $2 \div 3$ km on the average, the computation time is approx.10 s. In case of a single-iteration computation the deviations reduce to values below 1 km, while the computation duration increases roughly two times.

– the proposed methodological approach may serve as a basis for solving the problem of determining optimal maneuvers and orbital parameters correction within the set limits.

5. Optimization of hierarchic control system structure

Analysis of the results of variation problems solution relating to different hierarchy levels has shown the necessity for their complex conjugation and investigation of the problems of optimizing the used control hardware composition. It is explained by the fact that the results of solving the lower level problem considerably affect the input data for higher level problems. Thus the frequency of SC orbit altitudes at which spacecraft constellations operate affect the intensity of spacecraft avoidance maneuvers from space debris and corrections for maintenance of orbital parameters. Besides the orbit determination accuracy and SC motion prediction determine the precision of SC entry and landing in the predetermined landing site. Secondly, SC orbit injection, motion parameters determination and specification make it necessary to continuously use the control facilities. In case of large-scale deployment of orbit constellations this requirement would be more stringent thus necessitating the investigation of the optimization problem of control system utilization.

Now describe the general methodological approach to investigation of optimization problem of the hierarchic control system for the control of SC orbit constellations. In this work endeavors are undertaken to formulate and study the general objective of SC orbit constellation control using the waiting theory methods. These methods are widely used for investigating dynamic systems with the restricted structure [1-14], however they were not used rather fully for optimizing the structure of SC hierarchic control systems.

At first we state the methodological approach substance applying a concrete case of investigating the SC disturbed motion trajectory in atmosphere. In case of a SC atmospheric flight it is difficult to precisely take into consideration external disturbing factors since they are either unknown or of a complex character. Such external factors may include atmosphere density variations, wind, air turbulent motion, control action response and measurement errors, etc. The stated circumstances result in the necessity of reviewing the problems under uncertainty conditions.

To solve the problem the mathematical tool of the continuous Markovian process theory [17, 19] was used which is a general waiting theory component. The algorithms for computing mathematical expectations of reentry vehicle terminal parameters without mass computations of the disturbed trajectories were developed thus considerably reducing the time for quantifying the SC landing precision.

The algorithms development was based on reduction from the determinate description of phase point motion to the stochastic one [28].

The determinate presentation of SC motion in atmosphere is described by the differential equation system the latter being the system special case (3.1):

$$f_1 = \frac{dx_1}{dt} = \frac{dV}{dt} = -\frac{C_x S \rho V^2}{2m} - g_E \sin\theta,$$

$$f_2 = \frac{dx_2}{dt} = \frac{d\theta}{dt} = \frac{C_y S \rho V^2}{2m}\cos\gamma - \frac{g_E}{V}\cos\theta + \frac{V}{r}\cos\theta,$$

$$f_3 = \frac{dx_3}{dt} = \frac{d\varepsilon}{dt} = \frac{C_y S \rho V \sin\gamma}{2m \cos\theta} - \frac{V}{r}\cos\theta\cos\varepsilon\, tg\varphi, \qquad (6.1)$$

$$f_4 = \frac{dx_4}{dt} = \frac{dh}{dt} = V\sin\theta, \qquad f_5 = \frac{dx_5}{dt} = \frac{dL}{dt} = \frac{VR\cos\theta\cos\varepsilon}{r}\cdot\frac{1}{\cos\varphi},$$

$$f_6 = \frac{dx_6}{dt} = \frac{dL_\sigma}{dt} = \frac{VR}{r}\cos\theta\,\sin\varepsilon,$$

$$\varphi = L_\sigma/R, \quad g_E = \mu/r^2, \quad \rho = \rho_0 \exp(-\beta h), \quad r = R + h.$$

Where: ρ_0 — near earth surface atmosphere density, β — logarithmic coefficient of atmosphere density change as a function of height. The remaining variables are similar to those given in chapter 3.

Believe that the SC motion is presented as nonstationary Poisson event flows consisting of SC sequential transitions from one state to another [28]. This would enable to use the mathematical formalism of the Markovian process theory.

It is assumed that SC may be in the terminal state to be determined by a preliminarily selected algorithm of hierarchic phase space discretization. Each state is characterized by hexadimensional vector $i_k (k = 1, ..., 6)$ of phase coordinates. SC presence in each state is determined by the probability $P(i_k)$. Hence

$$\sum_{i_k}^{J_1} ... \sum_{i_6}^{J_6} P(i_k) = 1.$$

Having the known values (i_k), the statistical expectation of SC phase coordinates may be computed by the formula:

$$M[x_k] = \sum_{i_1=1}^{J_1} x_k(i_1) \sum_{i_2=1}^{J_2} \sum_{i_3=1}^{J_3} \sum_{i_4=1}^{J_4} \sum_{i_5=1}^{J_5} \sum_{i_6=1}^{J_6} P(i_k), \qquad (6.2)$$

where: J_k- number of hierarchy levels on k-axis of the hexadimensional phase space.

When the probabilities $P(i_k)$ change the mathematical expectation vector $M[x_k]$ obviously changes too; thus modeling SC motion.

According to the Poisson distribution law the probability of a random event (here the probability of a transition to the prescribed state per a unit time) is determined by the formula

$$P(t) = \lambda(t)e^{-\lambda(t)},$$

The Poisson intensity shall be taken equal to the rated values of differential equation right sides (6.1) in the fixed phase states:

$$\lambda(t) = \sum_{k=1}^{6} \frac{|f_k|}{\Delta x_k},$$

where Δx_k - decomposition interval along k-axis.

Write the formula as an example for determining the flow intensity carrying the spacecraft from state (2, 1, 2, 1, 2, 2) to state (2, 1, 3, 1, 2, 2), i.e. the transition when the azimuth angle ε changes:

$$\lambda_3(2,1,2,1,2,2) = \frac{1}{\Delta \varepsilon} \left| \frac{C_y S \rho_0 \exp(-\beta h(1)) V(2) \sin \gamma}{2m \cos \theta (1)} - \right.$$
$$\left. - \frac{V(2) \cos \theta \cos \varepsilon(2) tg\varphi(2)}{r(1)} \right|.$$

To determine the probability of SC presence in each of the given states one should derive a Kholmogorov differential equation system. The general equation for random state with the coordinates i_k has the form of

$$\frac{d\,P(i_k)}{dt} = -P(i_k)\sum_{k=1}^{6} \lambda_k(i_k) + \sum_{j=1}^{6} P\big(i_{k,k\neq j}, i_j - 1\big)\lambda_j\big(i_{k,k\neq j}, i_j - 1\big). \qquad (6.3)$$

Hence, if $i_j = 1$ (the state is initial along j - coordinate), then $\lambda_j(i_{k,k\neq j}, i_j - 1)$ will be equal to zero. Finally we would get a common linear differential equation system with constant coefficients of dimensionality

$$N = \prod_{k=1}^{6} J_k.$$

Thus the SC spatial motion in atmosphere described by differential equations (6.1) may be presented stochastically, i.e. as a system composed of N linear differential equations for transition probabilities (6.3) and analytical dependences (6.2) for computing the mathematical expectation vector of phase coordinates.

Now discuss the presentation of a spacecraft subject to random disturbing factor effects. We shall use the stochastic version as the basic one the main advantage of which over the determinate version consists in a principal opportunity for taking into consideration the sums characterizing the random effects in the differential equation right sides.

The general view of the SC initial differential equation system is as follows:

$$x = c + G(x,t)\xi(t), \tag{6.4}$$

where x, u – phase variable and control vectors, $G(x,t)$ – matrix sizing $k \times k$ (in the case under consideration $k = 6$).

To present the processes described by the system (6.4) using the Markovian chain we would make use of the Fokker-Plank-Kholmogorov equation (FPK-equation) [28]:

$$\frac{\partial \rho}{\partial t} = -\sum_j \frac{\partial}{\partial x_j}(a_j \rho) + \frac{1}{2}\sum_{j,k} \frac{\partial}{\partial x_j \partial x_k}(b_{jk}\rho).$$

Here ρ – probability density. The coefficients $a(x,u,t)$ and $b(x,t)$ are numerical (non-random) functions. The coefficients $a(x,u,t)$, $b(x,t)$ physically characterize the function change rate $x(t)$ and conventional dispersion of the random function $f(t)$ accordingly. The coefficients $a_j(x,u,t)$ and $b_{jk}(x,t)$ are found by the equations:

$$a_j(x,u,t) = f_j(x,u,t), \qquad b_{jk}(x,t) = \sum_{i=1}^{6} g_{ji}(x,t) g_{ik}(x,t).$$

Discretize the phase space as it is done for the disturbance-free version. The Kholmogorov general equation for a random state with the coordinates i_k ($k = 1, \dots, 6$) will be written as follows:

$$\frac{d\,P(i_k)}{dt} = -P(i_k)\sum_{k=1}^{6} \lambda_k(i_k) + \mu_k(i_k) +$$

$$+ \sum_{j=1}^{6} P(i_{k,k\neq j}, i_j - 1)[\lambda_j(i_{k,k\neq j}, i_j - 1) + \mu_j(i_{k,k\neq j}, i_j - 1) \tag{6.5}$$

with $\lambda_j = 1$ $\lambda_j(i_{k,k\neq j}, i_j - 1) = 0$.

The μ_j values may be interpreted as instantaneous values of flows carrying the spacecraft to neighboring states as a result of random function effects. Generally μ_j is calculated by the formula

$$\mu_j = \frac{1}{2}\sum_{k=1}^{6} \frac{b_{jk}}{\Delta x_i \Delta x_i}.$$

In particular, if the initial matrix $G(x, t)$ is diagonal, then the values of μ_j are determined by the dependence

$$\mu_j = \frac{1}{2} \frac{g_{jj}^2}{\Delta x_j^2}.$$

It is quite obvious that with $\mu_j = 0$ (no disturbances) the general equation (6.5) coincides with the equation (6.3).

So while integrating the N system of differential equations of the type (6.5) and applying the obtained probability values $P(i_k)$ for computing mathematical expectations (6.2) it is possible to determine SC motion trajectory in atmosphere with due consideration of random disturbing effects with unknown statistic characteristics.

When using the proposed computation technique there appear complexities associated with the large dimensions of the obtained differential equation system. For instance, even with a rather large discretization – decomposition of each coordinate axis into five intervals – we would get 15625 equations the numerical solution of which will be accompanied by an extremely great length of the computation process. Consider some ways of its reduction.

Transform only the first three equations of the system (6.1) in the stochastic form (6.1) – SC motion dynamics equations and the variables the changes of which are described by kinematic relations (h, L, L_σ) will be determined by approximated analytical formulae depending on the parameters V, θ, ε [29]:

$$h = -\beta \ln \frac{\theta^2 + A_2}{A_1 \rho_0},$$

$$L = L_0 + \frac{1}{\beta} \sum_{n=1}^{l} (-1)^{n+1} \left[\sum_{i=1}^{n} C_{in} \left(\theta_0^i - \theta^i \right) + C_n' A_3 + C_n'' A_4 \right], \qquad (6.6)$$

$$L_\sigma = L_{\sigma 0} + \frac{1}{\beta} \sum_{n=1}^{l+1} (-1)^{n+1} \left[\sum_{i=1}^{n} D_{in} \left(\theta_0^i - \theta^i \right) + D_n' A_3 + c A_4 \right],$$

where

$$A_1 = \frac{2}{\beta} \left(M - \frac{C_y S}{2m} \cos \gamma \right), \quad A_2 = \frac{2\rho_0}{\beta} \left(M - \frac{C_y S}{2m} \cos \gamma \right) - \theta_0^2,$$

$$M = \frac{1}{pr} \left(\frac{g_E r}{V^2} - 1 \right), \qquad A_4 = \text{arctg} \frac{\theta_0}{\sqrt{A_2}} - \text{arctg} \frac{\theta}{\sqrt{A_2}},$$

$C_{in}, C_n', C_n'', D_{in}, D_n', D_n''$ – piecewise constant coefficients, the algorithm for their solution is given in work [29], l – number of neglected terms in the trigonometric dependences decomposition $\sin \varepsilon$ and $\cos \varepsilon$ into the Maclaurin series.

52

The data computed applying the relations (6.6) for $t \leq 15$s contain certain errors as compared with the results of numerical integration of the equation system (6.1) not exceeding approx.3% [29].

Besides, with a rather small step of the Kholmogorov equation integration ($t \leq 15$ s) for computing the parameters h, L, L_σ the use of simple recurrence relations may be effective:

$$h_{i+1} = h_i + V_i \sin \theta_i \, \Delta t,$$

$$L_{\sigma \, i+1} = L_{\sigma \, i} + \frac{V_i R}{r} \cos \theta_i \sin \varepsilon_i \, \Delta t, \qquad L_{i+1} = L_i + \frac{V_i R \, \cos \theta_i \sin \varepsilon_i}{r \, \cos(L_{\sigma \, i}/R)} \Delta t.$$

Another way for reducing the computation burden lies in the fact that it is necessary to review not all phase space states obtained as a result of its discretization but only those the SC presence probability in which exceeds the given value P^*, as well as the states neighboring them. Thus initially one may be limited by three states along the velocity coordinates (V_1, V_2, V_3), trajectory ($\theta_1, \theta_2, \theta_3$) and azimuth ($\varepsilon_1, \varepsilon_2, \varepsilon_3$) angles. Hence $V_1 = V_0 + \Delta V$, $V_2 = 0$, $V_3 = V_0 - \Delta V$. Similarly $\theta_1 = \theta_0 + \Delta \theta$, $\theta_2 = 0$, $\theta_3 = \theta_0 - \Delta \theta$ and $\varepsilon_1 = \varepsilon_0 + \Delta \varepsilon$, $\varepsilon_2 = 0$, $\varepsilon_3 = \varepsilon_0 - \Delta \varepsilon$, i.e. initially the Kholmogorov equation system of 27 equations with the initial conditions $P_{222} = 1$ and $P_{ijk} = 0$ is integrated (for all cases but $i = j = k = 2$). When the probability $P(i, j, k)$ in an extreme state S_{ijk} reaches the value P^*, three states are added to which the spacecraft may be transformed directly from the S_{ijk} state, while the probabilities of SC presence at the given moment in the new states are taken equal to zero. And vice versa those states for which the probability becomes less than P^* are no longer considered (for further integration the Kholmogorov equations for these probabilities are withdrawn from the system).

The substance of the third way for reducing the computation length consists in a periodic updating of the initial conditions and phase space states used for Kholomogorov equations derivation. To avoid an excessive growth of the integrated equation system dimensionality at a certain time t^* the values of velocity mathematical expectations $M[V(t^*)]$ and angles $M[\theta(t^*)]$ and $M[\varepsilon(t^*)]$ computed by the formula (6.2) are accepted as new initial values of the phase coordinates V_0, θ_0, ε_0. In this case the t^* updating times may be either strictly set or determined by obtaining the prescribed number of Kholmogorov equations.

Though after introduction of the described updating the computation burden will be considerably reduced none the less it remains still rather great. Therefore another rather effective way for computation length reduction is used – analytical solution of the Kholmogorov differential equation system.

Derive a general formula enabling to compute the probability of SC presence in the state S_{ijk}. It is obvious that the spacecraft may get into the given state moving along different routes the total number of which is found according to the relation $n = f(i, j, k)$. According to the either-or theorem

$$P(i, j, k) = \sum_{r=1}^{n} P^r(i, j, k),$$

where $P^r(i, j, k)$ – probability of SC presence in the state S_{ijk} when moving along the r-route $(r = \overline{1, n})$.

Prove that with the predetermined route the probability $P^r(i, j, k)$ is found by the formula:

$$P^r(i, j, k) = B \sum_{n=1}^{m} e^{-k_n t} \Big/ \sum_{\alpha \neq n} (k_\alpha - k_n), \qquad (6.7)$$

where B – product of intensities of SC sequential transitions from the initial state to S_{ijk} state, m – number of states the spacecraft passes through moving along the predetermined route, k_s – sum of intensities of SC transitions from s-state $(s = \overline{1, m})$.

Give the evidence of the formula (6.7) validity applying the mathematical deduction. At first consider the case when $j = k = 1$. From one side the introduced concretization results in the convenience of mathematical statement since in this case there is no need to specify the SC route (it is determined explicitly), while on the other side it doesn't violate the generality of evidence, because the below-stated mathematical computations may be used for any predetermined SC motion route. First we shall prove the equality (6.7) for $i = 1$ and $i = 2$. The Kholmogorov equations for the states S_{111} and S_{211} have the form of:

$$\frac{dP(1,1,1)}{dt} = -P(1,1,1)k_1,$$

$$\frac{dP(2,1,1)}{dt} = \widetilde{\lambda_1}(1,1,1)P(1,1,1) - k_2 P(2,1,1),$$

where

$$\widetilde{\lambda_1}(1,1,1) = \lambda_1(1,1,1) + \mu_1(1,1,1), \quad k_1 = \sum_{j=1}^{3} \widetilde{\lambda_j}(1,1,1), \quad k_2 = \sum_{j=1}^{3} \widetilde{\lambda_j}(2,1,1).$$

Applying the Laplace operator method we solve these equations:

$$P(1,1,1) = e^{-k_1 t}, \quad P(2,1,1) = \widetilde{\lambda}_1(1,1,1)\left(\frac{e^{-k_1 t}}{k_2 - k_1} + \frac{e^{-k_2 t}}{k_1 - k_2}\right).$$

It is quite clear that the two solutions are special cases of a general solution (6.7). Further on condition that the equality (6.7) is valid for a certain value of i we shall prove its validity at $i + 1$.

Assuming that

$$P(i,1,1) = \prod_{n=1}^{i-1} \widetilde{\lambda}_1(n,1,1) \sum_{n=1}^{i} e^{-k_n t} \Big/ \prod_{\alpha \neq n}(k_\alpha - k_n). \tag{6.8}$$

We write the Kholmogorov differential equation for the probability $P(i + 1,1,1)$:

$$\frac{dP(i+1,1,1)}{dt} = \widetilde{\lambda}_1(i,1,1)P(i,1,1) - k_{i+1}P(i+1,1,1).$$

Applying the method of random constant variation we solve the equation:

$$P(i+1,1,1) = e^{-k_{i+1}t}\widetilde{\lambda}_1(i,1,1)\left[\int P(i,1,1)e^{k_{i+1}t} + C_{i+1}\right].$$

With due consideration of the relation (6.8) the last formula may be transformed in the following way:

$$P(i+1,1,1) = \prod_{n=1}^{i} \lambda_1(n,1,1)\left[\sum_{n=1}^{i} e^{-k_n t} \Big/ \prod_{\substack{\alpha=1 \\ \alpha \neq n}}^{i+1}(k_\alpha - k_n) + c_{i+1}e^{-k_{i+1}t}\right]. \tag{6.9}$$

The integration constant c_{i+1} is found on condition that $P(i+1,1,1) = 0$ with $t = 0$:

$$c_{i+1} = \left[\prod_{\substack{\alpha,n=1 \\ \alpha \neq n}}^{i+1}(k_\alpha - k_n)\right]^{-1}.$$

Thus the acquired dependence for probability $P(i+1,1,1)$ computing (6.9) satisfies the general solution (6.7), proving its validity.

The final form for determining the SC presence probability in state S_{ijk}, which can be entered along one of the routes may be presented as:

$$P(i,j,k) = \left[\prod_{s=1}^{m-1} \tilde{\lambda}(S) \sum_{s=1}^{m} e^{-k_s t} \Big/ \prod_{\substack{\alpha=1 \\ \alpha \neq s}}^{m}(k_\alpha - k_s)\right].$$

Make a comparative analysis of the results obtained applying the proposed method assuming use of all listed techniques for reducing the computation length and known numerical integration methods for differential equation systems (6.1).

55

The magnitudes of phase coordinate computation errors and SC motion trajectory computation length will be affected by the values of intervals $\Delta V, \Delta \theta, \Delta \varepsilon$, Kholmogorov equation integration step Δt, initial conditions updating interval t^* (or boundary probability P^*): decrease of $\Delta V, \Delta \theta, \Delta \varepsilon, \Delta t, P^*$ values and increase of t^* result in the computation errors reduction on one side and in the computation length increase on the other side. The numerical results analysis has shown that the most efficient method is application of the following constant values: $\Delta t = t^* = 1$ s. The $\Delta V, \Delta \theta$ and $\Delta \varepsilon$ intervals change while SC moves in atmosphere with 0.1 to 2-3 km/s, 0.1 to 4-6°, 0.1 to 2-4° accordingly, gaining maximum (minimum) values on legs where the rate of V, θ, ε variables is highest (lowest).

On the whole it was demonstrated that within approx.75–80% of the total duration of SC motion trajectories the quantitative differences of data obtained by the (6.1) equations integration from the results computed applying the developed algorithm do not exceed approx.1 - 3%. The computation length reduces by about 5 times as compared with the numerical integration methods.

The main problem restricting the use of these methods is that under real conditions there are no Markovian processes transforming a system from one state to another. All real processes are accompanied by one or another aftereffect almost without exception. Depending on the aftereffect degree the event flows are distributed according to other laws. The applicability of the waiting theory methods depends on the maximum system probability errors which may occur due to the formal substitution of real processes with Markovian processes (or real flows with Poisson flows).

The aftereffect method is known in the waiting theory [17], when real flows differing from the Poisson flows are approximated by the Erlang arrivals and certain fictitious pseudo-states are introduced in the pattern of probable system states. In this case the real HCS processes can be reduced to the Markovian processes and described using the Kholmogorov differential equation system. However one should note that application of the pseudo-state method for HCS studies is a poorly studied problem and constitutes a separate research discipline within the waiting theory.

Another principal feature of using the waiting theory methods for investigating the given problem is concretization of the HCS state elements. One should note that even under conditions of selected hierarchy levels such concretization is rather conditional. In association with the above-stated it seems expedient to use the approach based on selection of a number of alternative versions for shaping the system states with a subsequent efficiency assessment of each alternative version. A time minimum for the system to fulfill all its object functions may be selected as an efficiency criterion. As applied to the orbit constellation HCS it may be a time minimum for a space system formation with predetermined orbital characteristics and as applied to control of one-type spacecraft – a time minimum for fulfilling the object information acquisition program at customers' requests.

Now solve the problem of SC orbit constellation hierarchic control optimization. As an example we shall determine the system state space elements on the basis of the given rough individual element interaction pattern of a four-level hierarchic system.

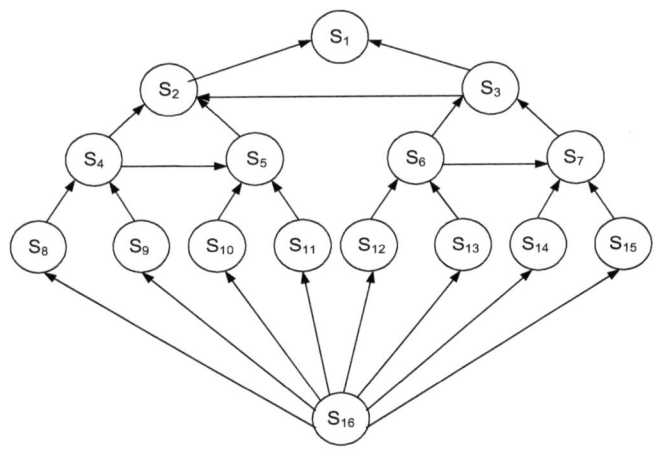

Interaction pattern of separate system components

Describe the state graph elements according to the depicted figure:

S_1 – mission profile fulfillment, efficiency assessment of control system configuration version;

S_2 – SC control optimization maintaining the orbital parameters within the set limits;

S_3 – SC control optimization during deorbiting, descent and earth landing maneuvers;

S_4 – ground instrumentation hardware control;

S_5 – SC orbit parameters determination, ballistic information computation;

S_6 – control of SC onboard instrumentation hardware for and space debris relative position determination;

S_7 – identification, determination and specification of space debris orbital parameters, generation of recommendations on SC avoidance maneuvers;

S_8 – SC motion parameters assessment after orbit injection;

S_9 – assessment of propellant consumption for SC orbit injection;

S_{10} – SC orbit correction, setting of required values of SC constellation orbital parameters;

S_{11} – optimal engine thrust vector control on powered flight leg;

S_{12} – processing of relay satellite performance information;

S_{13} – reception of relay satellite service requests;

S_{14} – processing of ground control facilities state information;

S_{15} – reception of ground facility service requests;

S_{16} – initial state.

In this case the system transitions from S_{16} to S_8, S_9, S_{10}, S_{11}, S_{12}, S_{13}, S_{14}, S_{15} states correspond to functioning of the elements of the fourth level of system hierarchy (selection level). The subsequent system transitions to S_4, S_5, S_6, S_7 states correspond to the third level elements functioning (adaptation level); transitions to S_2 and S_3 states – to the second level elements functioning (self-organization level); transitions to S_1 state – to the higher level functioning (coordination level).

The patterns depicted in the figure may be interpreted as a hierarchic structure system state graph. The system functioning process starts from S_{16} state and passing through S_i intermediate states terminates with the mission profile realization (S_1 state)

Designate the system presence probability P_i ($i = 1, \ldots, 16$) in S_i state. It is clear that at the initial time moment t_0:

$$P_{16}(t_0) = 1, \quad P_i(t_0) = 0 \quad \text{with} \quad i \neq 16. \tag{6.10}$$

For the presented state graph the probability values P_i are found solving the Kholmogorov differential equation system:

$$\frac{dP_1}{dt} = \lambda_2^1 P_2 + \lambda_3^1 P_3 , \quad \frac{dP_2}{dt} = -\lambda_2^1 P_2 + \lambda_3^2 P_3 + \lambda_4^2 P_4 + \lambda_5^2 P_5 ,$$

$$\frac{dP_3}{dt} = -\lambda_3^1 P_3 + \lambda_3^2 P_3 + \lambda_6^3 P_6 + \lambda_7^3 P_7 ,$$

$$\frac{dP_4}{dt} = -\lambda_4^2 P_4 - \lambda_4^5 P_4 + \lambda_8^4 P_8 + \lambda_9^4 P_9 , \quad \frac{dP_5}{dt} = -\lambda_5^2 P_5 + \lambda_4^5 P_4 + \lambda_{10}^5 P_{10} + \lambda_{11}^5 P_{11} ,$$

58

$$\frac{dP_6}{dt} = -\lambda_6^3 P_6 - \lambda_6^7 P_6 + \lambda_{12}^6 P_{12} + \lambda_{13}^6 P_{13} \,,$$

$$\frac{dP_7}{dt} = -\lambda_7^3 P_7 + \lambda_6^7 P_6 + \lambda_{14}^7 P_{14} + \lambda_{15}^7 P_{15} \,,$$

$$\frac{dP_8}{dt} = -\lambda_8^4 P_8 + \lambda_{16}^8 P_{16} \,, \qquad \frac{dP_9}{dt} = -\lambda_9^4 P_9 + \lambda_{16}^9 P_{16} \,,$$

$$\frac{dP_{10}}{dt} = -\lambda_{10}^5 P_{10} + \lambda_{16}^{10} P_{16} \,, \qquad \frac{dP_{11}}{dt} = -\lambda_{11}^5 P_{11} + \lambda_{16}^{11} P_{16} \,,$$

$$\frac{dP_{12}}{dt} = -\lambda_{12}^6 P_{12} + \lambda_{16}^{12} P_{16} \,, \qquad \frac{dP_{13}}{dt} = -\lambda_{13}^6 P_{13} + \lambda_{16}^{13} P_{16} \,,$$

$$\frac{dP_{14}}{dt} = -\lambda_{14}^7 P_{14} + \lambda_{16}^{14} P_{16} \,, \qquad \frac{dP_{15}}{dt} = -\lambda_{15}^7 P_{15} + \lambda_{16}^{15} P_{16} \,,$$

$$\frac{dP_{16}}{dt} = -\lambda_{16}^8 P_8 - \lambda_{16}^9 P_9 - \lambda_{16}^{10} P_{10} - \lambda_{16}^{11} P_{11} -$$
$$-\lambda_{16}^{12} P_{12} - \lambda_{16}^{13} P_{13} - \lambda_{16}^{14} P_{14} - \lambda_{16}^{15} P_{15} \,. \qquad (6.11)$$

The event flow intensities λ_j^i are determined as a rate of operations realization necessary for carrying the system from i to state j. Note that λ_j^i should be determined with due regard to effects of random disturbing factors which may result in a considerable change of the operations realization dynamics: technical facilities faults, actuator malfunctions, control operator's errors, etc.

We shall use the minimum time for carrying the system to state S_1 as a criterion of the variations problem solution, i.e.:

$$T_\Sigma = min \quad \text{with} \quad P_1(T) = 1 \quad \text{and} \quad P_i(T) = 0, \quad i = 2, 3, \ldots, 16. \qquad (6.12)$$

We state the variations problem with due consideration of the above-said: it is necessary to determine the set of transition intensities λ_j^i supporting the system transition from its initial state (6.10) to the terminal state (6.12) for the differential equation system (6.11).

The stated problem may be solved using the necessary optimality conditions [18]. Now we write the Hamiltonian

$$H = \sum_{i=1}^{16} P_i \psi_i$$

and conjugate variables

$$\frac{d\psi_1}{dt} = 0, \quad \frac{d\psi_2}{dt} = -\lambda_2^1 \psi_1 + \lambda_2^1 \psi_2 \,, \quad \frac{d\psi_3}{dt} = -\lambda_3^1 \psi_1 - \lambda_3^2 \psi_2 + \lambda_3^1 \psi_3 + \lambda_3^2 \psi_3 \,,$$

$$\frac{d\psi_4}{dt} = -\lambda_4^2 \psi_2 + \lambda_4^2 \psi_4 + \lambda_4^5 \psi_4 - \lambda_4^5 \psi_5 \,, \qquad \frac{d\psi_5}{dt} = -\lambda_5^2 \psi_2 + \lambda_5^2 \psi_5 \,,$$

$$\frac{d\psi_6}{dt} = -\lambda_6^3\psi_3 + \lambda_6^3\psi_6 + \lambda_6^7\psi_6 - \lambda_6^7\psi_7, \qquad \frac{d\psi_7}{dt} = -\lambda_7^3\psi_3 + \lambda_7^3\psi_7,$$

$$\frac{d\psi_8}{dt} = -\lambda_8^4\psi_4 + \lambda_8^4\psi_8 + \lambda_{16}^8\psi_{16}, \qquad \frac{d\psi_9}{dt} = -\lambda_9^4\psi_4 + \lambda_9^4\psi_9 + \lambda_{16}^9\psi_{16}, \qquad (6.13)$$

$$\frac{d\psi_{10}}{dt} = -\lambda_{10}^5\psi_5 + \lambda_{10}^5\psi_{10} + \lambda_{16}^{10}\psi_{16}, \qquad \frac{d\psi_{11}}{dt} = -\lambda_{11}^5\psi_5 + \lambda_{11}^5\psi_{11} + \lambda_{16}^{10}\psi_{16},$$

$$\frac{d\psi_{12}}{dt} = -\lambda_{12}^6\psi_6 + \lambda_{12}^6\psi_{12} + \lambda_{16}^{12}\psi_{16}, \qquad \frac{d\psi_{13}}{dt} = -\lambda_{13}^6\psi_6 + \lambda_{13}^6\psi_{13} + \lambda_{16}^{13}\psi_{16},$$

$$\frac{d\psi_{14}}{dt} = -\lambda_{14}^7\psi_7 + \lambda_{14}^7\psi_{14} + \lambda_{16}^{14}\psi_{16}, \qquad \frac{d\psi_{15}}{dt} = -\lambda_{15}^7\psi_7 + \lambda_{15}^7\psi_{15} + \lambda_{16}^{15}\psi_{16},$$

$$\frac{d\psi_{16}}{dt} = -\lambda_{16}^8\psi_8 - \lambda_{16}^9\psi_9 - \lambda_{16}^{10}\psi_{10} - \lambda_{16}^{11}\psi_{11} - $$

$$-\lambda_{16}^{12}\psi_{12} - \lambda_{16}^{13}\psi_{13} - \lambda_{16}^{14}\psi_{14} - \lambda_{16}^{15}\psi_{15}.$$

We obtain the equality from the transversability condition:

$$\psi_1(T) = -1. \qquad (6.14)$$

Taking into account the fact, that $d\psi_1/dt = 0$, we write:

$$\psi_1(T) \equiv -1.$$

The rest of conjugated variable values in the terminal trajectory point are found by solving the boundary value problem obtained by analogy with work [21] applying the method of successive approximations.

The analysis of relations (6.13) shows that when the system operates optimally all variables λ_j^i take their boundary values. In general terms the equation for determining intensity values of all transitions has the following form:

$$\lambda_j^i P_j(\psi_i - \psi_j) \to max$$

or

$$\lambda_j^i = \begin{cases} \lambda_j^i \, max & \text{with } \psi_i > \psi_j \\ \lambda_j^i \, min & \text{with } \psi_i < \psi_j \end{cases}.$$

Solution of boundary problems for the equation systems (6.11) and (6.13), boundary conditions (6.12), (6.14) enables to assess the special effect of variables λ_j^i for time T_Σ. It is demonstrated that all transition intensities λ_j^i may be divided into two categories. The first category includes such λ_j^i, whose value increase results in a considerable time decrease T_Σ. Concerning the example under review this is $\lambda_2^1, \lambda_3^1,$ $\lambda_4^5, \lambda_5^2, \lambda_6^7, \lambda_7^3, \lambda_8^4, \lambda_{12}^6, \lambda_{16}^8, \lambda_{16}^{10}, \lambda_{16}^{12}$. Variations of the remaining 14 values λ_j^i: $\lambda_3^2,$ $\lambda_4^2, \lambda_6^3, \lambda_9^4, \lambda_{10}^5, \lambda_{11}^5, \lambda_{13}^6, \lambda_{14}^7, \lambda_{15}^7, \lambda_{16}^9, \lambda_{16}^{11}, \lambda_{16}^{13}, \lambda_{16}^{14}, \lambda_{16}^{15}$ – practically have no effect on the total duration of object problems solution T_Σ.

It should be noted that improvement of promptness of solving the noted first-order problems may be assured both by changing their fulfillment intensity and introducing corrections in the hierarchic system structural configuration on the basis of efficiency assessment of various alternative versions.

6. Conclusions

The presented results enable to make the following principal conclusions:

1. The basic methodology principles for hierarchic control systems optimization of large SC orbit constellations have been developed and potentially the number of constellation spacecraft may reach several hundreds. At that not only the comparable assessment of the efficiency of using alternative versions is made but also determination of critical system elements for each version. It enables to introduce step-by-step corrections in the hierarchic system structural configuration.

2. The main hierarchy levels of space system functioning have been identified. For optimal control of space system elements in the initial space mission phase (orbit constellation formation) an analytical method for computing SC quasi-optimal motion trajectories on a powered flight leg for target orbit injection has been developed. The minimum energy consumption was taken as an optimality criterion. The main academic novelty of the described method lies in the developed algorithms for conventional powered flight leg partition into separate sections, identification of original special solutions and a subsequent conjugation of the results obtained. The method makes it possible to compute SC flight trajectories close to optimal and the selected orbit injection pattern ensures the propellant saving in weight equal to about 2.5-3 t in comparison with the known analogs when the SC launch weight is approx. 1500 t.

3. The critical problem occurring after SC orbit constellation formation is SC flight safety in man-made space debris environment. One of the effective methods for its solution is precise prediction of close space debris and SC approaches by onboard hardware and fulfillment of avoidance maneuvers when the collision probability is high. The work proposes the method for determining the orbital parameters of space debris motion using onboard optical sensors. The method novelty assuring a high level of computation accuracy and promptness of problem solution is transformation of the initial mathematical model of relative motion of objects and development of noniterative analytical algorithms. The conditions under which the needed accuracy of space debris motion coordinates computation have been identified. The given trend of SC mission safety enhancement is most promising because there are small-size man-made space debris fragments detection capabilities available and practically there are no ground-based capabilities for their tracking.

4. While orbital SC constellations are used as target oriented systems often it becomes necessary to fulfill dynamic operations (risk object avoidance maneuvers, orbit parameters correction, deorbiting). The method for determining the optimal engine thrust vector control was developed for determining optimal SC deorbiting maneuvers and SC transfer to the predetermined phase space region applying the Pontryagin maximum principle. High efficiency of the SC control applying the two-impulse method was demonstrated. In certain cases the propellant mass saving as compared with the ordinary single-impulse pattern constitutes 10%. The given solution may serve as a basis for investigating the generally stated problem of reactive thrust control optimization.

5. SC atmosphere flight dynamics were studied. The noniterative algorithm for computing the ballistic descent and landing in the predetermined earth surface area was developed. Use of this algorithm enables to assure a high SC landing accuracy (with deviation from the predetermined landing point of 2-3 km maximum). The methodological novelty of the algorithm is the original approximated analytical solution of the SC motion differential equation system serving as a basis of the algorithm.

6. Analysis of the SC atmosphere disturbed motion dynamics shows that to precisely take into account the effect of disturbance forces is extremely difficult because they are either unknown or of a complex nature. To solve the problem under disturbance force uncertainty an algorithm for computing the stochastic characteristics of reentry vehicle terminal motion parameters was developed. The algorithm was based on reduction from the motion deterministic to stochastic description, derivation and analytical integration of the differential equation system for computing SC presence probability in the given phase space region. It was shown that concerning 75-80% of the total SC motion trajectory length the quantitative differences of the data obtained applying the known numerical integration methods and the data obtained applying the developed algorithm does not exceed 1-3%. In this case the computation time reduces by about 5 times as compared with the numerical integration procedure application.

7. The process of identifying the structure configuration optimization of the hierarchic control system is described applying the specific example of its presentation. For the identified hierarchy levels the main state space elements were determined, the structure elements interaction pattern was given, the Kholmogorov differential equations were derived for computing the system presence probability in

a predetermined state, the dependences for determining the optimal system operating conditions were found. The general methodological approach to problem solution of SC orbit constellation hierarchic control optimization may find a wide application in many areas of scientific research and applied activities.

List of references

1. Moiseev N.N. Information theory of hierarchical systems. Materials of all-Union seminar on large systems control. Moscow. 1983.

2. Denisov A.A., Kolesnikov D.N. The theory of large-scale control systems. L.: Energoizdat. 1982.

3. Mesarovich M., Takahara Y. Theory of hierarchical multilevel systems. M.: Mir. 1973.

4. Tzvirkun A.D. The structure of complex systems. M.: Soviet Radio. 1973.

5. Erlikh A.I., Ven V.L. Structure choice of programming process. Programming method of control. N.: Computation Centre of the Russian Academy of sciences. Vol.3. 1973.

6. Singh M., Titli A. Systems: decomposition, optimization and control systems. M.: Mashinostrojenye. 1986.

7. Kulikowski R. Optimization of large-scale system. Automatica. 1970.

8. Kulikowski R. Modelling of production, utility structure, process and technological change. Control and Cybernetics. 1975.

9. Guberinic S. On structural optimization in hierarchical system. Prepins of the IFACIFORS symposium. Verna. 1974.

10. Bronstein O.I., Tzvirkun A.D. About hierarchical control systems. Automatics and telematics. №1. 1968.

11. Vlasyuk V.A., Morosanov I.S. Design of hierarchical structure of control in large systems. Automatics and telematics. №3.1973.

12. Chunchenpak I.M. The structure optimization for resource allocation systems. Materials of the institute of electronic control machines. M.: 1976.

13. Voronin A.A., Mishin S.P. Optimal hierarchical structures. M.: Institute of control sciences, Russian Academy of sciences. 2003.

14. Bagdasaryan A.G. A general structure of information expert system for simulation and analysis of complex hierarchical systems in control loop. Control of large-scale systems. M.: Institute of control sciences, Russian Academy of sciences. №3. 2008.

15. Sokolov N.L., Pozniak T.I., Pinchuk V.B. The use of the theory of Markov processes to study spacecraft motions in the atmosphere. Cosmic research. Vol. 28, №3.1990.

16. Zagorulko A.M. Integrated automated control system for space vehicle. Data processing systems. №2. 2008.

17. Ventzel E.S. Operations research. M.: Soviet radio. 1982.

18. Pontryagin L.S., Boltyanskiy V.P., Gamkrelidze R.V., Mishchenko E.F. Mathematical theory of optimal processes. M.: Science. 1969.

19. Bellman R. Dynamic Programming. M.: Foreign literature. 1960.

20. Anorov V.P. Maximum principle for processes with general constraints. Automatics and telemechanics. №3,4. 1967.

21. Ivanov N.M., Martynov A.I. Movements of spacecraft in atmospheres of planets. M.: Science. 1985.

22. Andrievsky V.V. Dynamics of spacecraft descent. M.: Mashinostroyenie. 1970.

23. Sikharulidze Y.G. Ballistics of space vehicles. M.: Science. 1982.

24. Ivanov N.M., Dmitrievsky A.A., Lysenko L.N. Ballistics and navigation of spacecrafts. M.: Mashinostroyenie. 1986.

25. Bettin R. Guidance in space. M., Mashinostroyeniye. 1966.

26. Elyasberg, P. E. Introduction to the theory of flight of artificial earth satellites. M.: Science. 1965.

27. Letov A.M. Flight dynamics and control. M.: Science. 1969.

28. Sveshnikov A.A. Applied Methods of the theory of random functions. M.: Science. 1968.

29. Sokolov N.L. Approximate analytical method for calculation of space maneuvers in the atmosphere. Cosmic research. №2. 1988.